AGRICULTURAL AND RURAL RECONSTRUCTION

Dr. Jayant Patil was born in Bordi, Maharashtra in March 1927. He earned M.Sc. (Agriculture) Degree from University of Poona. Under Fulbright Programme he went to United States and obtained M. S. Degree from Kansas State University. The Konkan Agricultural University conferred Doctor of Science on him. After his higher education in India and U.S.A. he chose to spend most of his working life of over 40 years in remote areas of Thane District working with the tribals who constitute weakest section of society in India. Hunger, malnutrition and under employment were the main problems of these tribals. Dr. Patil developed appropriate technologies for sustainable development. He has developed social approaches for transfer of these technologies. The utilisation of these technologies have brought about food security and employment generation among the tribals. Dr. Patil had visited several countries to study agriculture and rural development. He tried to introduce various agricultural technologies suitable under Indian conditions. Dr. Patil received Jamnalal Bajaj Award in 1979 for application of Science and Technology for Rural Development. The Government of India appointed him to the prestigious post of Member, Planning Commission in charge of Agricutlure, Rural Development, Irrigation, Panchayati Raj and Cooperation in August 1991.

AGRICULTURAL AND RURAL RECONSTRUCTION

A Sustainable Approach

DR. JAYANT PATIL

(Member, Planning Commission)

CONCEPT PUBLISHING COMPANY, NEW DELHI - 110059

ISBN 81-7022-589-2

First Published 1996

Published and Printed by
Ashok Kumar Mittal
Concept Publishing Company
A/15-16, Commercial Block, Mohan Garden,
New Delhi - 110059 (India).
Phones : 5554042, 5504042

FOREWORD

Dr. Lester Brown, President of the World Watch Institute, in a recent book titled "Who will feed China?", states "the earth's capacity to produce enough food to satisfy our expanding demand is now emerging as the overriding environmental issue as the world approaches the twenty-first century. The loss of food security promises to become the defining focus of the global environmental threat". This book, based on Dr. Jayant Patil's life long work in tribal and rural areas, is thus a timely contribution, since it shows how sustainable food and nutrition security can be achieved through environmentally sustainable and socially equitable development. Dr. Patil has brought out clearly through his own work how it is essential to integrate the principles of ecology and equity with those of economics, if the concept of sustainable development has to be converted from rhetoric to reality.

The book is written in a simple, narrative style and therefore makes a deep impact. Dr. Patil describes how from the time he passed his B.Sc. Agriculture degree from the College of Agriculture, Pune, in 1949, until January, 1983 when he left the Kosbad Institute, he had an exciting adventure in the field sustainable and equitable development of tribal areas. The uniqueness of the Kosbad model of sustainable development is its human-centered approach to natural resources management. The *Kosbad model* of sustainable agriculture and rural development spearheaded by Dr. Jayant Patil involves a holistic approach to the improvement of the quality of life of tribal and rural children, women and men. Dr. Patil has demostrated how *antyodaya* (i.e. attention to the poorest of the poor) has to be the pathway to achieve a *Sarvodaya* (where public good and private well being become mutually reinforcing) society.

The book provides a glimpse of the work done at Kosbad over a period of 34 years to produce more food, more jobs and more income through concurrent attention to farm and off-farm employment. Dr. Patil is a firm believer in the principle that if the human resource is undervalued and if material resources are overvalued, poverty cannot be eliminated. He therefore followed the Gandhian dictum that to achieve rural regeneration and agrarian prosperity, we need an appropriate blend of brain and brawn. He established not only a Research Institute but also a *Krishi Vigyan Kendra* where tribal women and men can learn new technical skills through the method of learning by doing. Thus, Dr. Patil helped to convert know-how into do how.

Dr. Patil's work both at Kosbad and subsequently in various capacities have been chronicled in a lucid manner in this book. The credibility of his message is high, since it is based on personal experience. Dr. Patil has shown how poverty alleviation can be achieved through the technological and information empowerment of the poor. His methods involve the substitution of knowledge for capital and farm grown biological inputs for market-purchased chemical inputs. In other words, Dr. Patil brings out clearly through this little book that there is a tide of new possibilities for achieving a hunger - and poverty free India. As Shakespeare wrote,

> "there is a tide in the affairs of Men which taken at the flood leads on to fortune."

As Dr. Ismail Seregeldin, Senior Vice-President of the World Bank said recently, we need fortune not just in terms of more economic growth in monetary terms, but

> "fortune in terms of true well being of all,
> fortune in terms of quality of life
> fortune in terms of leaving for our children and
> our children's children a better world"

This book indicates a pathway for achieving this type of fortune. It should hence be read by all students and faculty of Agricultural and Rural University as well as by the general public and administrative and political leaders.

(M. S. SWAMINATHAN)
President
National Academy of Agricultural Sciences and
the National Foundation for India

Madras

December 24, 1995

PREFACE

I used to attend the evening prayers of Mahatma Gandhi while I was studying at the college of Agriculture in Pune. This prayer meeting was held at the Nature Cure Clinic of Dr. Dinshaw Mehta which is situated near the Pune Railway Station. Gandhiji used to address the audience after the prayers. The subjects included Hindu-Muslim Unity, Service of Scheduled Castes and Tribes, Women's Organisations, Khadi and Village Industries. Gandhiji's language was very simple. He used to speak with utmost sincerity. His every word would therefore touch our hearts. Once he spoke of Thakkar Bappa's exemplary service of tribals. Thakkar Bappa sacrificed all wordly comforts and went to a remote place called Dohad in the Panchmahal District of Gujarat. He dedicated his life to the service of the Bhil tribals of that area. Gandhiji said that we were still in the need of social workers like Thakkar Bappa. The educated youths from our country should devote themselves to the service of tribals. At that moment, I resolved that I should also work among the tribals.

The village Bordi in Thane district where I was born was surrounded by a large tribal population. I observed the poverty of these people from my childhood. My father used to tell me about the exploitation of tribals by forest contractors, landlords and traders. During pre-independence days the forest contractors were harassing the tribal labourers. Acharya Bhise under whose guidance I received my secondary education, dedicated his life to the service of tribals and so I was brought up in the atmosphere of service of tribals right from my school days. When I passed my Bachelor of Science (Agriculture) examination from the college of Agriculture, Pune, Acharya Bhise expressed a desire that I should work at the Agricultural Institute at Kosbad Hill* and devote myself to the cause of Rural Reconstruction and Tribal Welfare. This Institute was just started by him. Respecting his wish, I joined "the Kosbad Institute" on June 20, 1949. I worked there upto January, 1983 for a period of about 34 years. During this period I received an opportunity to experiment in Rural Reconstruction.

---

* (The name of this Institute will appear frequently throughout this book. For the sake of convenience I shall refer to the Agricultural Institute at Kosbad Hill, simply as "the Kosbad Institute".)

Fortunately, I got opportunities to visit the U.S., Netherlands, England, Australia, Israel and Philippines. I could study Agriculture and Rural Development in those countries. On my return from these countries, I tried to introduce, with suitable modifications, agricultural technologies and rural development aspects which I learnt there.

We have to adopt an integrated approach in Rural Reconstruction, so I did not restrict my experiments only to agriculture but included other areas like education, health and village industries.

The most important principle in rural reconstruction is the development of the individual. Rural reconstruction will not be real if the sons of the soil are disturbed. We have to recognize that the sons of the soil have the first right on the area where they live. If we forget their natural rights and carry out developmental activities, it would simply be economic development. Ignoring the human aspect would lead to a very serious problem of refugees. If such a thing happens, it would not be real development.

Mahatma Gandhi had given very basic thinking about rural reconstruction fifty to sixty years back. His thoughts are relevant even today. He used to say that *creation of wealth is important,* but with the important rider *for whom*? The economics of morality and immorality is based on the reply to this question.

It is therefore necessary to make the lives of rural people more comfortable and tolerable. The real meaning of rural reconstruction is to give the rural people economic strength through their daily programmes, to enable them to march ahead for the welfare of the future generations and so to develop the dormant qualities in them. The main objective of my experiments in rural reconstruction is to prepare human beings who will be able to meet the challenges of modern times according to Gandhian thinking.

Today, most of the rural reconstruction work is done by the Government departments, although Voluntary Agencies are also engaged in rural reconstruction. Government departments have to follow a schematic pattern in carrying out rural reconstruction programmes. While voluntary organisations have the freedom to make necessary changes in the schematic pattern, this flexibility is not available to the Government departments. Besides, the working among the tribals require a humanitarian approach. The Kosbad Institute was working among the tribals and so we had to be flexible and humanitarian in implementing the rural reconstruction programmes. I have made these experiments in this context.

During my stay in the United States, I have seen scientists who selected only one subject and did research on it for all their lives. They did not leave their laboratories during their lifetimes. I received great inspiration from such dedication to the cause of science. I was also convinced that the rural reconstruction programmes would not work unless they are strengthened by science and technology. Therefore, when I returned to India, after completing my education in the United States, I started the research wing in the Kosbad Institute. The experiments which we did in the laboratory strengthened the rural reconstruction programmes. The Kosbad Institute, in a real sense, served as a laboratory of rural reconstruction work.

I was very fortunate to receive guidance from saintly persons having great wisdom. Their advice was very useful to me in my experiments in rural reconstruction. The distinguished persons both from India and abroad who came to see experiments gave me great encouragement. The tribals, who are very innocent and straight forward, willingly cooperated, giving me strength in my rural reconstruction work.

The rural reconstruction experiments which I carried out around Kosbad Hill are also being done by other voluntary agencies in their areas. There is a need to establish closer linkages of these experiments to help strengthen the rural reconstruction movement in India. With this object in view, I present this book.

JAYANT PATIL

ACKNOWLEDGEMENT

When I conceived the idea of bringing out my experiences in a form of a book, I could see a number of problems and limitations. The initial work was of difficult nature but as I proceeded, I received valuable guidance and support from a number of persons who deserve special mention.

Dr. M. S. Swaminathan, eminent agricultural scientist of international repute and presently President, National Academy of Agricultural Sciences and the National Foundation for India, provided me moral as well as material support which greatly helped in finalising my thoughts into writings. I am also grateful to him for writing the Foreword to this book.

I am grateful to the staff of Kosbad Institute for their valuable assistance in my work for all these years.

Shri G. S. Guha, Incharge, Agro-climatic Regional Planning Unit at Ahmedabad deserve my thanks for his help in editing and organising the computer print-out of this book. I also thank Shri O.P. Grover, Senior Consultant, Shri M. D. Narula, Private Secretary and Smt. Rachna Bawa in Planning Commission for their valuable assistance. My special thanks are due to Shri Ashok Kumar Mittal of Concept Publishing Company, who very much cooperated in bringing out this book in time.

JAYANT PATIL

CONTENTS

1

INSTITUTIONS OF AGRICULTURE AND RURAL DEVELOPMENT

Establishment of the Agricultural Institute at Kosbad Hill

The Department of Agriculture, Bombay State had drawn up a plan in 1947 to start one Agricultural School in each district. The objective of starting these schools was to give two years of agricultural training to the sons of farmers, so that, after completion of training, they would go back to their lands and start adopting new technologies. All these schools were to be run by the Agricultural Department of the Bombay State. Only one exception was made, that of Thane District.

Acharya Bhise, a great educationist, had started Agricultural Education in his S.P.H. High School at Bordi in 1928. At that time nobody had ever thought of starting agricultural education in the rural schools. The first Chief Minister of Bombay State Shri Balasaheb Kher had personally seen the agriculture education work started by Acharya Bhise and was greatly impressed with it, so he insisted that the proposed agriculture school in Thane District should be started under the guidance of Acharya Bhise. The Government of Bombay, therefore, sanctioned a scheme to start the agricultural school at Bordi. The services of Shri Harishchandra Patil, an agricultural expert, were made available to conduct the school. Shri Patil had obtained B.Ag. degree from the College of Agriculture at Poona in 1932. He had devoted himself to the cause of agriculture and rural development. Since that time he had vast experience in this field. The Gokhale Education Society, a premier educational organization in the state, accepted the responsibility of management of the school. The State Government, for the first time, gave the responsibility of managing an agricultural school to a non-government organization.

The school was first started at Bordi. However, the Bordi campus proved inadequate for agricultural training. So Acharya Bhise and Shri

Harishchandra Patil selected a site at Kosbad Hill and requested the State Government to acquire this land. This land had an area of 265 acres. On 40 acres, mango, chikoo, guava and coconut trees were planted. There was a big well with sufficient water for irrigation. There was an oil engine fitted on this well connected with necessary pipelines to irrigate the fruit trees. Besides, there were residential buildings and cattle-sheds. Shri Krishnaswamy Iyengar had purchased this property in 1940 with a noble objective of starting a Hanuman Service Centre (Sevashram). So he planted fruit trees and started a dairy farm. But he had to sell the property due to ill-health. Later on, this property was purchased by Shri Kamlakar Deshpande from Devas of Madhya Pradesh. However, he died there and so the land was on the name of his wife Mrs. Hemlatha Deshpande. During this time, there was an agitation among the tribals of Kosbad Hill and surrounding areas under the auspices of the Communist Party of India. The Police had to resort to firing on the mob who were opposing the bullock-carts loaded with grass. However, nobody was injured in this firing. Due to this agitation, no labour was available for the orchards and so the fruit trees suffered. Besides, there was no *pucca* road for coming to Kosbad. One had to come in bullock-cart. There were two rivulets between Gholvad and Kosbad and so during floods the place was completely isolated. The officers of the State Agriculture Department who came to make the survey of this property recommended to the Government that this property was not suitable for running the agriculture school as it had no infrastructure like roads and communication. The Bombay Government was, therefore, reluctant to acquire this property for the agriculture school.

Shri Annasaheb Vartak, a great social worker and a leader of Thane District was at that time a Minister in the Government of Bombay. When he came to know that there was an adverse report on the property at Kosbad, he personally came for inspection. There was no *pucca* road for going to Kosbad and so Shri Vartak had to travel in a horse driven *tonga*. He visited the property with Acharya Bhise and Shri Harishchandra Patil. The land was situated in a valley between two hills. There was good resources of ground water. Shri Vartak said that due to availability of ground water the land was suitable for growing fruit trees and vegetables and would be most suitable to give training in horticulture and vegetable cultivation to the sons of the farmers. Besides, the land was situated in the tribal area and so it would help to give agricultural training to the tribals. This would be an ideal centre for the welfare of the tribals. Kosbad Hill overlooks the Arabian Sea and the site was very beautiful. Shri Vartak made a strong recommendation to the Government of Bombay that this property should be acquired for starting the Agricultural School. He also convinced his colleagues in the Cabinet. The Government of Bombay ultimately decided to acquire 265 acres of land at Kosbad. In this way the efforts of Shri Annasaheb Vartak were responsible for starting the Agricultural School at Kosbad. The responsibility of acquiring this land was entrusted to Shri Khanderao Save, who was the Deputy Collector at Dahanu. Shri Save had his education under the guidance of Acharya Bhise at Bordi and so he had the ideology of service to the tribals. He, therefore, carried out the acquisition procedure speedily and the property was acquired in April, 1948. After the acquisition of this property, the responsibility of management was entrusted to the Gokhale Education Society. Shri Pandoba Patil, a practical farmer from Chinchani was appointed to manage the fruit orchard. He took great pains to restore the orchard which was neglected for the last several years. He was constantly moving in the garden taking care of every plant, till he became disabled.

The students and staff of Agricultural School at Bordi, shifted to Kosbad. However, the students did not have a hostel and the staff also did not have quarters. Similarly there was no communication due to lack of all weather road. There was urgent need of construction of road and buildings. Shri Vaikunthlal Mehta, then Finance Minister of Government of Bombay suggested to Acharya Bhise that the land and property at Kosbad be entrusted to the Government. Similarly the management of the agricultural school may also be transferred to the Agricultural Department of the State Government. This would facilitate construction of road and buildings and also development of a training farm. Accordingly, the Gokhale Education Society handed over the management of the agricultural school at Kosbad on 1st April, 1949 to the Agricultural Department of the Government of Bombay. The Government desired that the services of Shri Harishchandra Patil be placed at the disposal of Government of Bombay for running the Agricultural School. The Gokhale Education Society gave consent to Shri Patil to work as the Principal of the Government Agricultural School. Similarly, the other staff members, Shri Raghunath Savale, Shri Narayan Patil and Shri Ramchandra Raut were also deputed to the Government Agricultural School.

I passed my B.Sc. (Agriculture) from the College of Agriculture, Poona in May 1949. I went to meet Acharya Bhise in the Shardashram at Bordi. Acharya Bhise told me, "We have started an Agricultural School at Kosbad and it is my desire that you join this Institute and work there. However, you first meet Shri Harishchandra Patil, then go to Kosbad and then decide". The next day, I went to Kosbad with Shri Harishchandra Patil, who took me around the farm and showed the various activities. He also expressed the desire that I should join this Institute.

On my return from Kosbad, I told my father, "This Institution is situated in the tribal area. The tribals are following primitive methods of cultivation and so their crop yields are very low. They have to undergo hunger for a part of the year. If we give them training in improved practices of agriculture their crop yields would improve and their hunger could be eliminated. I could utilise my knowledge for educating the tribals in new agricultural technology". My father was very happy on hearing my words. He told me, "The tribals in our area are very poor and it is our duty to serve them. You will be able to utilise your agricultural education for this very useful purpose. Besides, you will get the guidance of Acharya Bhise which will shape your life". I accepted the advice of my father and accordingly I joined the Agricultural School at Kosbad on 20th June, 1949.

There were no facilities at Kosbad at that time. Since there was no road we had to walk 5 miles from Bordi. During monsoon, when the streams were flooded, all communication came to a halt. If somebody fell sick, we had to carry him in the bullock cart to the dispensary at Gholvad. Food articles and other groceries were brought on head-loads. There was only one old building in which staff and students shared their accommodation. Shri Harishchandra Patil, the Principal of the School was also staying with the staff as no other accommodation was available.

Shri Vasudevrao Chitre, Shri Narayan Patil and myself were staying together in one room. Shri Raghunath Savale and Shri Ramchandra Raut were residing with their families in a small cottage. The kitchen was established in huts prepared from bamboo and rice straw. During heavy rains, these huts leaked. However, the staff and the students gladly shared all the inconvenience caused to them. Acharya Bhise used to come regularly and enquire about our difficulties. He was trying his best to solve the problems as quickly as possible. One day he came and told us that the approach road to Kosbad of 1-1/2 mile had been sanctioned by the Government. The forest department was given the responsibility for its construction. At that time, there was a Training Centre at Bordi for training of the foresters. Shri Abhyankar was the Principal of that Centre. He was entrusted with the responsibility of constructing the road. He was a very enthusiastic officer. He immediately started construction of this road and within two months he completed the earth work. Thereafter, gravel was spread over it. Two bridges were also constructed on this road. Shri Bhausaheb Hirey, Minister for Revenue, Agriculture and Forests, Government of Bombay, inaugurated these bridges in September, 1952. In this way, the isolated village of Kosbad was connected by road with Bordi and Gholvad villages.

The Government of Bombay also sanctioned the construction of buildings of the Agricultural School at Kosbad. The Public Works Department of the Government was entrusted with the work of the construction of these buildings. Shri R.K. Patil, Food Commissioner, Government of India laid the foundation stone of the building of the Agricultural School in March, 1950. Shri Balasaheb Kher, Chief Minister of Bombay State inaugurated this building in April, 1951. Shri M.P. Patil, Minister for Agriculture, Shri Annasaheb Vartak, Minister of Local Self Government were present on this occasion. Large number of tribals and other farmers from Thane district attended this function. The starting of the Agricultural School at Kosbad had opened a new vista in the field of agricultural development and tribal welfare.

Transfer of Government Agricultural Institute to Voluntary Agency

The director of the Kosbad Institute, Shri Harishchandra Patil, was appointed by Government of Bombay as Officer on Special Duty for Japanese Method of Rice Cultivation for Bombay State in 1952. He therefore, went to Pune which was the headquarter of the Directorate of Agriculture of Bombay State. Since no new appointment took place for Director of the Kosbad Institute I was officiating as its Director for about a year.

I had arranged a rice demonstration and invited Shri Yeswantrao Chavan who was then the Civil Supplies Minister of Bombay State. We planted rice by Japanese method of rice cultivation. Shri Chavan was greatly impressed to see the rice crop. He gave us great compliments for the efforts we were doing to increase the food production, as the country was short of food at that time. Since that time Shri Chavan has known the Kosbad Institute. Whenever I used to meet him in Bombay or New Delhi he was enquiring about the activities of the Institute.

Afterwards Shri A.B. Joshi was appointed as Director of the Kosbad Institute. During his tenure one new well was constructed on the farm. He also established a floricultural garden in front of the office building. Shri Joshi stayed for a year and was then appointed District Agricultural Officer at Dhule. Thereafter Shri V.H. Joshi was appointed as Director. A number of rice demonstrations were arranged during his tenure. Eminent personalities like Dr. Panjabrao Deshmukh, Union Minister of State for Agriculture, Shri V.B. Eshwaran, Secretary to the Government of Bombay, Agriculture Department and Dr. T.G. Shriname, Director of Agriculture, Bombay State attended these demonstrations.

In 1955, Shri Harishchandra Patil took a decision to return to his original area of work at Kosbad Hill. Accordingly he tendered his resignation from the post of officer on Special Duty. Shri Bhausaheb Hiray was the Minister for Agriculture at that time. Acharya Bhise and Shri Harishchandra Patil explained to him that the activities of the Kosbad Institute should not be restricted simply to imparting agricultural training to the sons of farmers but its sphere of activities should be widened so that all the tribals and small farmers in the areas could be benefitted. It would be necessary to start an educational institution for tribal children as well as training institutions for rural development functionaries. They requested him to entrust the responsibility of running the Institute to Gokhale Education Society which was a voluntary agency. My father Shri Shamrao Patil also convinced Shri Hiray about the need of widening the activities of Agricultural Institute in the interest of tribals and weaker sections. My father was the General Secretary of Maharashtra Pradesh Congress Committee when Shri Hiray was its President. Their relations were therefore very cordial.

Shri Bhausaheb Hiray was convinced about the need of widening the activities of the Kosbad Institute. Accordingly the Government of Bombay passed a resolution in December 1955 to transfer the management of the Government Agricultural Institute to the Gokhale Education Society. The land, buildings and equipment were also transferred to the Society. This shows the vision of Shri Bhausaheb Hiray about the role that voluntary organisations could play for promotion of agricultural development among the weaker sections.

When the management of the Government Agricultural Institute was handed over to the Gokhale Education Society the Departmental staff of the Institute was transferred elsewhere in the State. Acharya Bhise and Shri Harishchandra Patil desired that I should continue to work in the Institute. They requested the Government of Bombay to send me on deputation to the Gokhale Education Society. The Government granted their request and accordingly I continued my work at the Institute.

When Shri P.K. Sawant was Agriculture Minister he transferred property of the Government Agricultural School to the Gokhale Education Society. This has enabled the Society to erect new buildings and make additions like wells and other structures in the agricultural farm. When the Kosbad Institute was run by the Government only one programme of imparting agricultural training to the sons of farmers was carried out and so the activities of the Institute had remained very limited. When the Kosbad Institute was given to the voluntary agency, Gokhale Education Society, it received guidance of dedicated workers like Acharya Bhise and Shri Harishchandra Patil. The Institute, therefore, was able to work towards education and development of weaker sections like small farmers and tribals. The Institute had undertaken an integrated programme for the educational development of tribal children which included Creche, *Balvadi* (pre school), *Ashram* School (residential primary school) and Secondary School. The Institute also started training rural reconstruction workers at centres like Panchayat Raj Training Centre. This has helped to create extension workers for rural reconstruction work.

The Kosbad Institute also started the Mahatma Gandhi Adivasi Janata Vidyalaya for imparting practical training to the tribal youths. The Krishi Vigyan Kendra (Farm Science Centre) was also started to upgrade skills of farmers and fishermen. A research wing was started where research on soils and crops was undertaken. The Kosbad Institute adopted the surrounding villages for transfer of agricultural technology in food crops, oilseeds, horticulture, animal husbandry, fisheries etc. This shows that if a voluntary institution is given opportunity it can make significant contribution to research, education and extension.

Agricultural Training for Tribal Youths

An All India Tribal Welfare Conference was held at Bordi in 1959. Dr. Rajendra Prasad, President of India inaugurated this Conference. Shri Govind Ballabh Pant, Union Minister for Home, Government of India also attended this conference. At that time the tribal welfare department was under the Ministry of Home Affairs and so Acharya Bhise placed a scheme for agricultural training of tribal youths before Shri G.B. Pant. The main objective of this scheme was to train the tribal youths in modern agricultural technology so that they could go back to their lands and improve the productivity of their agriculture.

Shri G.B. Pant immediately approved this Scheme and instructed the officers for its speedy implementation. Shri D.R.Pradhan, Secretary of Social Welfare Department, Government of Bombay visited Kosbad Hill next day to see the site and work out other details. We proposed to name this Institute as " Mahatma Gandhi Adivasi Janata Vidyalaya."

It was proposed that the land adjoining the Kosbad Institute be purchased for construction of buildings and setting up demonstration and training farm. It was also proposed to train fourty tribal youths at a time from six districts viz. Thane, Raigad, Nasik, Ahmednagar, Dhule and Jalgaon. It was decided to start a long-term course of one year and short-term courses of two to three months,and Sylabi were prepared for these courses. Emphasis was given on practical training.

Acharya Bhise and Shri Harishchandra Patil told me that since this institution would be the first of its kind in Bombay State, we should make maximum efforts to make it a success. They, therefore, desired that I should accept the responsibility of conducting this Institution. Accordingly, I accepted the post of Director, Mahatma Gandhi Adivasi Janata Vidyalaya.

A hilly land was purchased for construction of buildings. It was proposed to construct hostel buildings for trainees on this hill. The foundation stone of the hostel building was laid by Shri Laxmidas Shrikant, Commissioner for Scheduled Castes and Scheduled Tribes, Government of India. Shri Shrikant was born in a wealthy family in Bombay. Soon after graduation he went to Mahatma Gandhi to seek his guidance. Gandhiji told him to go to Dohad and meet Thakkar Bappa who was working among tribals. Shri Shrikant listened to the advice of Gandhiji and went to Dohad. He worked under the guidance of Thakkar Bappa and devoted his whole life for service of tribals. Dr. Rajendra Prasad, President of India recognised the services of Shri Shrikant and appointed him as the first Commissioner for Scheduled Castes and Scheduled Tribes of Government of India. While working in this high post he lead a simple life and made significant contribution in development of tribals in the country. It was, therefore, most appropriate that the foundation stone of the hostel was laid by a devoted person like Shri Shrikant.

There was a provision of a community hall in the scheme. We decided to name it the 'Balasaheb Kher Hall'. Late Shri Balasaheb Kher was the first Chief Minister of Bombay State and founder of Adivasi Sewa Mandal (Tribal Service Society). Shri Kher had deep concern for the upliftment of tribals. The foundation stone of the Balasaheb Kher Hall was laid by the great Sarvodaya Leader Shri Jayaprakash Narayan. His wife Shrimati Prabhavati Narayan was also present on the occasion. Shri Jayaprakash Narayan was greatly impressed by the scheme of imparting agricultural training to tribal youths. He told us that in Bihar there was a large population of Santhal tribals who needed such training. He sent workers from his Sukhodeora Ashram in Bihar to study this scheme.

Shri Jayaprakash Narayan also visited the horticultural orchards of tribals around Kosbad Hill. He was impressed to see that the tribals were getting gainful employment on their land. He said that such pattern of horticultural development would be needed for the whole country. This would bring about socio-economic development of weaker sections.

We also showed him the mango orchard on the farm of the Institute. He told me that in Sadakat Ashram at Patna there were a large number of mango trees of 'Malda' variety but they were not bearing fruits. He told me to visit the Ashram whenever I go to Patna. I used to go to Patna for meetings of the Rice Development Council. Accordingly I visited the

Sadakat Ashram. Shri Mrityunjaya Prasad, son of Dr. Rajendra Prasad took me round the ashram and showed the mango trees. All the trees were heavily infested with powdery mildew and jassid hoppers. It was very necessary to spray some insecticide and fungicide. After returning to Bombay I contacted Shri Lalloobhai Patel, Chairman of Aspee Foundation, Malad. Shri Patel was Chairman of a firm which manufactured plant protection equipment. I requested him to help the Ashram authorities in plant protection of mangoes. When I mentioned the name of Sadakat Ashram, he told me that he was a freedom fighter and was very proud of Sadakat Ashram which gave great inspiration to hundreds of freedom fighters like him. He immediately instructed his agent in Patna to carry out plant protection of mangoes in the Ashram. He also donated one power sprayer to the Ashram. During that year the mango trees bore good fruit. When Shri Jayaprakash Narayan came to know about this he wrote me a letter of appreciation. After a few days when I went to Delhi I met Jayaprakashji in Gandhi Peace Foundation. He again expressed his gratitude for the interest I took about the spraying of mango trees for the Sadakat Ashram. I told him "I did a very small duty, for which you need not thank me. Instead I seek your blessings".

The inauguration of Balasaheb Kher Hall was done by Shri Vaikunthlal Mehta, a great Gandhian and Chairman of Khadi and Village Industries Commission. Shri Mehta had done pioneering contribution in cooperative movement in Bombay State.

Shri R.S. Patankar a leading architect from Bombay had prepared the plans of the buildings. He did all the work without charging any professional fees. Whenever we requested him to charge some fees, he used to say that Acharya Bhise has done immense sacrifice for this Institution and so it was his duty to help it. Shri Manjibhai Lalji Patel from Dahanu was given the contract for construction of buildings. Acharya Bhise knew Shri Patel since 1947 when he undertook construction of buildings of Saryodaya Scheme in Kasa. Even though Kasa was in the interior forest areas and did not have any communication facilities, Shri Patel did a good job of the construction. Shri Patel constructed the buildings of Mahatma Gandhi Adivasi Janata Vidyalaya also in time.

A land measuring ten acres in Kosbad valley was purchased for setting up a training and demonstration farm. An irrigation well was constructed and fitted with a pump. The trainees planted four hundred chikoo grafts to celebrate Independence Day, on 15th August 1960. Now there is a beautiful orchard of *chikoos* on this farm.

Tribal youths from five districts were given training in agriculture. We gave emphasis on practical training which helped them to acquire skills in growing of food crops, oil seeds, vegetables, fruit trees, fodders and agro-forestry. We also gave them practical training in animal husbandry, dairying, fisheries, sericulture and bee keeping. After completion of training, every trainee was supplied with a sample of improved seeds, plants, and grafts which helped them to put training into practice. The trainees then adopted improved agricultural practices on their lands and increased their yields. Their farms proved to be demonstration centres for neighbouring farmers.

Shri Morarji Desai former Prime Minister of India was Chairman of Bharatiya Adim Jati Sevak Sangh (All India Tribal Service Organisation). He suggested to me to conduct training for tribal youths from other States also. He instructed Shri J.H. Chinchalkar, the then Secretary of the organisation, to depute youths from various centres in the country. Accordingly the tribal youths from all over India were coming for training at Kosbad Institute. We were giving them agricultural training which would be suitable to their agro-climatic conditions. Besides we used to show them various developmental activities like Forest Labour Cooperative Societies and Ashram Schools. They were also shown the agricultural development projects among the tribals which included multiple cropping, establishment of fruit orchards and nutrition gardens. The trainees were greatly inspired to see such self-help programmes. Thus the agricultural training to tribal youths helped to improve agricultural productivity of their lands, and created self confidence among them about tribal development.

Rural Development Training Centres

In the bigger bilingual Bombay State Shri Ratubhai Adani was the Minister for Village Panchayats and Sarvodya while my father Shri Shamrao Patil was the Deputy Minister of these departments. Once Shri Adani and my father visited the Kosbad Institute. We showed them the rural development activities which we were carrying out. They also saw the agricultural research and educational work of the Institute. They felt that the Institute was most ideal for giving training to functionaries of Village Panchayats as well as social workers from the Sarvodya Schemes. Accordingly they sanctioned two centres viz. Village Panchayat Training Centre and Sarvodaya Workers Training Centre, to be set up within the Kosbad Institute.

In the Village Panchayat Training Centre the Secretaries of Panchayats as well as *Talatis (Patwaris)* from Thane, Nasik and Raigad districts used to come for training. Shri Joshi and Shri Kulkarni who were in the rank of *tahsildars* were appointed by Government as training officers.

We used to organise camps for village sanitation. Shri Krishnadas Shah who devoted his life for village sanitation used to conduct these camps. Such camps were organised for the *talati* trainees in Gholvad village. The trainees had to get up early in the morning. Community prayers were organised which were also attended by villagers. The trainees had to work for eight hours to take various programmes like preparing compost pits, trench urinals, trench latrines, soak pits etc.

The local latrines were of the basket type. These baskets were carried by scavengers. Shri Krishnadas Shah told the trainees that this work of carrying human excreta is inhuman and should be discontinued. He wanted to convert these latrines into Gopuri latrines which did not require scavengers. Besides, the excreta could be transformed into manure. The participants gladly agreed to his proposal. They first cleaned the basket type of latrines and constructed Gopuri latrines in their place. The villagers of Gholvad held all these trainees in high esteem for this work. This shows that Government functionaries like *Patwaris* could do valuable rural reconstruction work if they come in contact with committed social workers like Shri Krishnadas Shah.

The Patwari trainees also offered their *shramdan* (voluntary labour) in digging the well of a tribal farmer Shri Wajya Dhak. They worked for about three hours every morning and completed the work in a month. One circle officer Shri Raut from Agashi continued this tradition of social work while in the service of the Revenue Department. He used to bring the farmers from his area to show the rural reconstruction work being carried out at Kosbad Hill. Thus the *Patwari* trainees who got opportunity to study in the Village Panchayat Training Centre received great inspiration to do social work.

In Bombay State, a Sarvodaya Scheme was started soon after Independence. In every district, a group of villages were selected for rural development work. All the functionaries in the scheme were workers of voluntary organisations. The work of the Sarvodaya Scheme was conducted on Gandhian lines. These workers were very committed but lacked in professional skills. We therefore gave them training in agriculture, animal husbandry, dairying and agro-forestry. These workers became more effective in their areas, after receiving skill training.

In the State of Maharashtra, the Panchayati Raj started in 1963 and so the Village Panchayat Training Centre was transformed into two centres viz. the Panchayati Raj Training Centre and Gram Sevak Training Centre. We used to impart training to the elected members of Village

Panchayats and also of Panchayat Samitis of blocks. We found that if the elected members of Panchayats are given orientation in agriculture, horticulture, agro-forestry, animal husbandry and dairying they develop insight for preparing village development plans. Similarly when the *gramsevikas* are given training in agriculture and allied subjects, they become more effective in transfer of agricultural technology in rural areas.

We found that trainees of the Rural Development Centres acquire good insights into the needs and priorities of people as also into methods of resolving them. Given opportunities, they are better equipped to participate in rural development programmes in villages.

Krishi Vigyan Kendra

The Indian Council of Agricultural Research had prepared a scheme to start Krishi Vigyan Kendras (Farm Science Centres) in the country. The objective of these Kendras was to upgrade the skills of farmers and fishermen through training. The theme of the training was 'learning by doing'. The first Krishi Vigyan Kendra in the country was started at Pondicherry in 1974.

Dr. M.S. Swaminathan, Director General, Indian Council of Agricultural Research during his first visit to Kosbad Institute in 1974 told us about the scheme of Krishi Vigyan Kendras. He desired that we should make an application to the Council through the State Government for granting a Krishi Vigyan Kendra at Kosbad. Accordingly, we approached Shri Vasantrao Naik, Chief Minister of Maharashtra and told him about this scheme. Shri Naik knew the tribal development work which we were doing at Kosbad Institute. He readily agreed to our request and recommended our application to Indian Council of Agricultural Research.

Shri Jagjivan Ram, Union Minister for Agriculture visited Kosbad Institute in January 1975. Shri Annasaheb Shinde, Minister of State for Agriculture, Government of India was also with him. We showed him the tribal development work done by the Institute. He was very happy to see the tribal women acquiring the skills of grafting mangoes and sapotas. One of the tribal farmers showed him his sapota orchard and told him that uptil now he was working as a labourer in orchards of rich farmers but now he had his own orchard. He now had year round gainful employment on his land. Shri Jagjivan Ram was so impressed by the work of the Kosbad Institute that he told in the meeting, "which could be the better place for Krishi Vigyan Kendra than Kosbad Institute?" Shri Annasaheb Shinde who was present on the occasion invited me to come to Delhi. Accordingly, I met Shri Annasaheb Shinde at Krishi Bhavan in Delhi. Shri Shinde called Shri O.P. Gautam, Deputy Director General, Indian Council of Agricultural Research and explained to him about the tribal uplift work done by Kosbad Institute. Dr. Gautam was impressed to hear this from Shri Annasaheb Shinde. He told him that he would immediately constitute a committee which would visit Kosbad Institute to study the possibilities of setting up Krishi Vigyan Kendra.

The Indian Council of Agricultural Research constituted a committee headed by Dr. J.S. Patel, former Vice-Chancellor of Jawaharlal Nehru Agricultural University at Jabalpur. Five senior scientists from Agricultural Universities and Central Institutes were members of this committee. The committee visited the Kosbad Institute and saw the infrastructure like, training and demonstration farm, hostel, class-room, laboratory, and library at the Kosbad Institute. They also visited the extension activities of the Institute in surrounding tribal villages. They were impressed with the work of Kosbad Institute and unanimously recommended that the Krishi Vigyan Kendra be started here. Accordingly, the Indian Council of Agricultural Research took the decision to start first Krishi Vigyan Kendra in Maharashtra at Kosbad Hill in 1976.

We designed the courses at Krishi Vigyan Kendra based on the needs of tribal farmers. Food was the basic need and so we trained them in cultivation of high yielding varieties of rice. During this course we invited the officers of the Panchayat Samiti (Block Council) to explain to the trainees various schemes in tribal sub-plan. This greatly helped the trainees to avail the benefits of schemes and put their training into practice. The tribals adopted high yielding varieties of rice and became self reliant in food production.

The tribals grew pulses and oilseeds on their uplands during rainy season. We organised training in cultivation of pulses and oilseeds. Scientists from Konkan Agricultural University at Dapoli were invited to give practical demonstrations. Similarly, the State Agricultural Department deputed its officers to explain about pulses and oilseeds development programmes. The trainees adopted improved practices in cultivation of pulses and oilseeds.

There was widespread problem of malnutrition in the area and so we organised a course in 'Nutrition Gardens' for tribal women. We gave them training in production and consumption of nutritive foods. Every tribal women established a nutrition garden in her backyard. The vegetables and fruits produced in this garden were included in their diet. We found that the self-help programme of 'Nutrition Gardens' helped the tribals to overcome problems of malnutrition.

Some of the tribal farmers had constructed small irrigation wells under the schemes of tribal sub-plan. We organised training for these farmers in cultivation of vegetables. The tribal farmers explained to us that due to lack of roads in the tribal region they found it difficult to transport vegetables in time. This put them to great losses. We, therefore, designed a course of growing non-perishable vegetables which consisted of onions, chillies, garlic, elephant yam, ashgourd, red pumpkin etc. After the completion of training the trainees introduced non-perishable vegetables on their lands. Growing such vegetables helped the tribals to get cash income thereby improving their purchasing power.

Most of the tribals have lands which are either rainfed or dry. We therefore, organised courses in dryland horticulture. Mangoes, cashews, custard apple, *ber, amla* and jackfruit were the fruit trees suitable for dryland horticulture. We also organised courses in irrigated horticulture for farmers who had wells. Sapotas, coconuts, lime, guavas, bananas and papayas were some of the fruit trees suitable under irrigated horticulture. The courses consisted of cultivation practices, plant propogation and post harvest practices. Training in horticulture has helped the tribal farmers to introduce fruit trees on their lands. This was the first generation of tribals to establish such fruit orchards.

The work animals of tribals were very weak and so they were unable to put in hard labour required for ploughing and puddling. The main reason for their weakness was lack of nutritive fodders. We therefore, arranged for training in cultivation of nutritive fodders viz., stylosanthes hamata, stylosanthes scabra and siratro. Since these fodders were new to them we supplied them with seeds and seedlings. This enabled them to put their training into practice. Introduction of nutritive fodders has helped to improve the productivity of cattle.

The tribal women have to undergo drudgery of carrying fuelwood from long distances. They collect firewood by cutting trees from forest areas, which leads to deforestation. We therefore, designed a course in agro-forestry for tribal women. The course contained raising of nursery of fuelwood trees, planting and care of trees, and improved *chulhas* (stoves). The tribal women were supplied with seeds and seedlings after the course which enabled them to cultivate fuelwood trees on their lands. This helped the tribal women to be self reliant in fuelwood.

The tribal villages are situated near forest area. The flora in the forest was congenial for bee-keeping. A course in bee-keeping was initiated both for tribal men and women. The district Village Industries Officer of Maharashtra State Khadi and Village Industries Board was invited to explain the scheme for supply of bee boxes and centrifugal machine for extraction of honey. After training the tribals started bee-keeping and obtained good incomes through the sale of honey. We found that bee-keeping was a good village industry for landless people.

The tribal areas have mango trees of local varieties. Besides there were other berries like *karondas (Carissa caranda)*. A course was designed in preparation of pickles from mangoes and berries. The tribal women who attended this course prepared pickles for the first time in their life.

The Indian Council of Agricultural Research organised a workshop of training organisers at Krishi Vigyan Kendras at the Kosbad Institute. Shri Bhanu Pratap Singh, Minister of State for Agriculture, Government of India inaugurated this workshop. In his inaugural address he said that the Krishi Vigyan Kendra at Kosbad is the most ideal centre in the country.

The training at Krishi Vigyan Kendras at Kosbad was helpful in upgrading the skills of tribal farmers both men and women. The tribals who were trained in cultivation of high yielding varieties of rice became self reliant in food production. The training in nutrition gardens enabled them to overcome problem of malnutrition. The tribals who received training in vegetable cultivation and horticulture developed good vegetable and fruit gardens. They received cash income through sale of vegetables and fruits, which enhanced their purchasing power. Significant improvement was recorded in their diet, clothing and housing. In this way the Krishi Vigyan Kendra played an important role in the upliftment of tribals through the transfer of technology.

Post Graduate Research Centre

I conducted research on rice under the guidance of Professor U.K. Kanitkar. The refuse of dried leaves and grass from the hills is washed down by the rain water and accumulate in rice fields. Due to excessive organic matter in the soil the growth of rice crops is very vigorous. This is locally called *Bhat Majne* (running to leaf). Excessive vegetative growth causes the rice crop to lodge on the ground and the filling of the grains do not take place which adversely affects the yields. The farmers therefore practice retransplanting of rice. This is locally called *Posha*. Some farmers cut the leaves with sickle. However due to cutting of leaves the earheads of rice become small and have less grains. The practice of *Posha* requires considerable amount of labour which is not available at that time. We therefore, conducted an experiment to prevent leafiness in rice. A Plant growth regulator viz. 2, 4-D was sprayed on the rice crop in different concentrations and different growth periods. The excessive vegetative growth of the rice crop was arrested by this chemical. The crop did not lodge on the ground and the yield was also good. I selected a rice variety called Bhades 12-11 for this experiment. I wrote a thesis on my research and submitted it to the University of Poona and I was awarded degree of Master of Science in first class.

After receiving my degree I met Prof. Kanitkar at Poona. He told me that we should try to get the Kosbad Institute recognized as a Post Graduate Research Centre. This would facilitate the scientists at the Institute to conduct research and obtain post graduate degrees. The Institute would also gain reputation through the research.

We made an application to the University of Poona for getting post graduate recognition. Availability of eminent Professors for guiding the post graduate students, facilities of laboratory, library and land for conducting research were the conditions required for such recognition. Prof. U.K. Kanitkar and Dr. S. Soloman agreed to work as Honorary Professors. We established a laboratory in an old building. We had been collecting laboratory equipment for the last 10 years. Hence,we did not require more expenditure for setting up the laboratory. A library was also established in the adjoining room. We were already subscribing to research journals and magazines, making it easier to set up the research library. The farm of the Agricultural Institute was very well laid out and was suitable for conducting research.

The University of Poona appointed a Committee of two members to examine the facilities at the Kosbad Institute. Dr. Mahabale, Head of the Botany Department, University of Poona was Chairman, while Dr. B.S. Kadam former Joint Director of Agriculture, Maharashtra was member of this Committee. The Committee visited the Institute and suggested to strengthen the facilities of library and laboratory. The Committee recommended post graduate recognition in agronomy, genetics and agricultural botany. Accordingly the University of Poona gave recognition to the Agricultural Institute, Kosbad as Post Graduate Research Centre. Thereafter, Agricultural Universities were established in Maharashtra and so the post graduation recognition was transferred to these Universities.

The centre was very fortunate to get the services of eminent scientists. Prof. U.K. Kanitkar was Professor of Botany at the Agricultural College at Poona. Besides, he had worked as Cotton Physiologist at Dharwar in Karnataka. He studied plant growth regulators at the University of Oxford. Dr. Samuel Solomon was a botanist and served as principal of the Agricultural College at Poona. He had an M.S. degree from the University of California. He conducted research on fodders.

After a few days Dr. N.R. Bhat joined the Institute as Post Graduate Professor. Dr. Bhat had obtained his Ph.D from the University of

Cambridge. He served as Director of Sugarcane Research Institute, Coimbatore and Tobacco Research Institute, Rajmahendri of Indian Council of Agricultural Research. Later we requested Dr. B.S. Kadam to work as Post-graduate Professor. Dr. Kadam was a well known plant geneticist. He had obtained his Ph.D in genetics from Cornell University in the U.S.A. He had evolved improved varieties in rice, wheat and gram. He had served as Director of Agriculture in Saurashtra State. Dr. R.D. Asana also agreed to join as Post-graduate Professor. Dr. Asana had served as the head of Plant Physiology Department at the Indian Agricultural Research Institute, New Delhi. He was staying at Bulsar after his retirement. He used to visit the Institute and conduct research on wheat.

Thereafter Dr. N.G. Magar who worked as head of the Department of Bio-chemistry at the Institute of Science, Bombay also joined the Kosbad Institute. I knew Dr. Magar since he had helped me to conduct soil analysis in the laboratory of the Institute of Science when I was conducting my rice research. We requested him to help us in setting up the Laboratory and guide the students in research. We were very fortunate to get the services of these eminent scientists who offered their services honorarily in guiding the students and also in building the Post-graduate Research Centre.

The Kosbad Institute was established in the year 1949. We celebrated the silver jubliee in the year 1974. Shri Vasantrao Naik, Chief Minister of Maharashtra was the chief guest on the occasion.

We submitted a project of Plant Science Laboratory to the Chief Minister. We told him that this project would enable the young scientists to conduct research under the agro-climatic conditions of Konkan. It would also strengthen the Post-graduate Research Centre of the Kosbad Institute.

Swami Anand who was present on the occasion told the Chief Minister that this laboratory would strengthen the research activities at the Institute. Shri Vasantrao Naik was a great visionary. He laid the foundation of scientific agriculture in Maharashtra. He could see the importance of plant science laboratry. He therefore, announced a grant of four lakhs of rupees for building of this laboratory.

The building was completed within two years. Necessary equipment was also secured during this period. The inauguration of this laboratory was done by Shri Shankarrao Chavan, Chief Minister of Maharashtra.

The laboratory was in need of good equipment. Dr. N.G. Magar prepared a project costing about six lakh rupees. Mr. Gunther Dheil, Ambassador of Federal Republic of Germany who visited the Kosbad

Institute was taken around the laboratory and shown the research work being carried out. We told him that if we got modern instruments the research would be strengthened. We also told him that these instruments were available in West Germany, and submitted a copy of the project report to him. After about one month we received a letter from the Ambassador inviting me to Delhi. I went to meet him at the Embassy of Federal Republic of Germany in Chanakyapuri at Delhi. He had also invited the representatives of a donor agency, Miserior from Germany. I told them about the agricultural development work that we were doing among the tribals of Kosbad Hill. I also explained to them about the need of equipment for strengthening the agricultural research work at the Kosbad Institute. Ambassador Dheil also supported our proposal for getting equipment for laboratory. The representatives of Miserior told me that they would send this proposal to their head office in Germany. Subsequently, we received a letter from Miserior from Germany, saying that they had sanctioned our project. They further stated that this equipment was being sent by ship to Bombay Port.

When this equipment reached Bombay Port we were informed by the customs department that a duty of rupees two lakhs would have to be paid. It was not possible for the Institute to pay such a large amount. We were informed that the Ministry of Finance, Government of India could waive the custom duty for imports meant for research purposes. Shri H.M. Patel was the Finance Minister at that time. However, I did not know him personally. I therefore went to Delhi to meet Shri Narendrabhai Nathwani who was Member of Parliament. Shri Nathwani used to come to Kosbad Hill to meet Swami Anand and he knew the activities of the Institute. I requested him to recommend our case to the Finance Minister to waive off the custom duty. Shri Nathwani met the Finance Minister and told him about the agricultural research work we were doing for the welfare of weaker sections. The Finance Minister considered our case and waived the custom duty. We brought this valuable equipment from the Bombay Port.

Dr. N.G. Magar sent one expert from Bombay who set up this equipment. Dr. N.R. Bhat and Dr. Magar prepared one more project of research on Pulses (grain legumes) and submitted to an organisation in Sweden. This organisation gifted us a valuable equipment called 'protein analyser'. There were many difficulties in importing this instrument. I approached Dr. O.P. Gautam, Director General, Indian Council of Agricultural Research who helped us. The plant science laboratory was greatly strengthened due to the equipment received from West Germany and Sweden.

We strengthened our library by subscribing to research journals and books both from India and abroad. I was sent to United Kingdom under the Colombo Plan to study British Agricultural Education System. On my return I received a letter from British Council that they would be presenting books worth one thousand pounds sterling. I went to the British Council Library with Dr. S. Soloman and prepared a list of scientific books. We received a very good collection of books through the grant.

The Post-Graduate Research Centre enabled the teacher scientists at the Institute to conduct research on food crops, vegetables, fruit trees and fodders. They submitted their thesis to the University and obtained Master of Science (Agriculture) degrees. We found that Post-graduate Research Centre was very useful for enabling the young scientist to tackle agricultural problems through research.

2

TRIBALS OF KOSBAD HILL

Land Reforms

The villages around Kosbad Institute are inhabited by tribal population. The tribals are the original inhabitants of this area. However, due to ignorance and poverty their lands were forcibly taken up by landlords and moneylenders and the tribals were rendered merely as tenant cultivators. The tribals had to pay half the share of their agricultural produce to the landlords, a practice under the local name of *Khand*. The tribals did not have any right on the trees existing on their lands. In the month of November when the threshing of rice was carried out the landlords demanded their share of the produce, without contributing their labour in the production process.

The tribal tenants were evicted by landlords according to their whims. With a view to protect the tenants from such evictions the first Congress Government passed a Tenancy Act in 1939. The tribal cultivators were declared as 'protected tenants'. After independence a Land Reform Act popularly known as 'Land to the Tiller' was passed in 1951 by which the tribals became the owners of their lands.

Many a times the lands alloted to tribals according to the tenancy act were situated in the midst of large holdings of landlords. In such cases the landlords with the help of local revenue officials indulged in changing the land records and forcing the tribals to surrender their lands. Often during night time boundaries of lands belonging to the tribals were bulldozed so that there would be no trace of their lands. Coercive methods were thus being used for such evictions. The landlords sold these big holdings to wealthy persons who wanted to establish fruit orchards. This was brought to the notice of Shri Shankarrao Chavan who was then the Chief Minister of Maharashtra during 1976. Shri Chavan visited the Kosbad Institute and heard the grievances of tribals whose lands were forcibly taken by landlords. He immediately instructed the high revenue and police officers present there to prosecute such landlords. We had arranged a big meeting of the tribals at Kosbad Institute where the Chief Minister declared that ".... Government would immediately pass an act in which a non-tribal cannot purchase the land of tribal. If any tribal wants to sell his land he will have to surrender it to the Government. The land transactions between non-tribals and tribals would be illegal". The remedial measures taken by Shri Shankarrao Chavan immediately put a curb on the landlords and land brokers from evicting the lands of tribals, and made justice available to the poor tribals of the area.

The moneylenders also exploited the tribals by advancing loans at high rates of interests. They used to recover loans by coercive means. The tribals were in great need of foodgrains during June and July which were the months of agricultural operations. The traders handed grain to the tribals on loan. After the harvest of rice crop in November the tribals had to return these grains twice or thrice in kind. This was locally known as *Palemod.* We brought this to the notice of Government, which reacted by passing a law banning the *Palemod* system. Grain depots were started in cooperatives through which grains could be had on loan at reasonable rates. However this scheme worked well only where there were good cooperatives.

I found that it required great courage and commitment to take up the work of removing exploitation. At Kosbad Institute we had committed workers who were constantly with the tribals and organized them to fight against injustice. One amongst them was Shri Bhikoo Sapte. Shri Sapte was a tribal youth whom we deputed for higher education at Konkan Agricultural University at Dapoli. After obtaining B.Sc. (Agri.) degree he returned back to Kosbad Institute to do agricultural development work among tribals. While doing this work he noticed the exploitation meted out to tribals by landlords and moneylenders. Shri Sapte took up the work to remove this injustice. He organized the tribals and took up their cases to the courts of law or to the concerned officials. Many a times he was threatened for his life by vested interests. He was very courageous and did not care for such threats. Due to his tireless efforts many tribals got justice in restoration of their lands. However, unfortunately Shri Sapte was killed in a bus accident and we lost a very committed worker. I have observed that government land reforms would become effective only if committed workers take up the work of organizing the weaker sections to fight against injustices.

Infrastructure Development

During pre-independence period and even after independence for many years the tribal areas did not have any road communication and medical facilities. I observed that many tribals died due to epidemics like small pox and cholera. Women died during pregnancy or delivery. Similarly due to snake bites many deaths were taking place. When we established the Kosbad Institute in 1949 we first undertook a programme of construction of roads. During fifties and early sixties this work was being done by the Thane District Local Board. After establishment of Thane Zila Parishad and starting of Employment Guarantee Scheme in Maharashtra, all the tribal hamlets were linked with roads. While making alignment of roads some landholders objected to spare lands for the proposed roads. Such landholders were approached personally by us and persuaded to surrender their lands voluntarily. We tried to avoid land acquisition procedures which took a long time in implementation. I found that due to such personal appeals the construction of roads in tribal areas was expedited. We made efforts to start the State Transport Buses linking Dahanu Road where medical facilities were available. In emergency cases we often spared vehicles of the Kosbad Institute to carry patients to the hospitals at Dahanu Road. I found that roads and communications played an important role in creating access to health facilities for the tribals.

Before independence the tribal areas did not have any educational facilities. Some primary schools were there in taluka towns. However, it was not possible for the boys to go from interior areas to attend these schools. The Adivasi Sewa Mandal under the guidance of Acharya Bhise started hostels for tribal boys in mid-forties.This enabled some tribal boys to attend primary schools. Soon after Independence, the Government of Bombay started a scheme to establish 'one teacher schools'. Government desired that the orientation of these teachers be done under the guidance of Acharya Bhise. These teachers were highly motivated and prepared to undergo all hardships while working in tribal areas. After the establishment of Kosbad Institute in 1949, we started one teacher schools in the neighbouring villages. These schools catered to the educational needs of tribal children upto the fourth grade. However, when the numbers of grades went on increasing it became difficult for one teacher to manage the entire school. These one teacher schools created awareness of primary education among the tribal children.

The Government of India prepared a scheme of Ashram schools in mid-fifties. This scheme was based on the pattern of residential schools where students and teachers could stay together. We started one Ashram

School at Kosbad Institute in 1957. Due to the availability of residential accommodation, tribal boys and girls from interior areas could attend the school. We found that the pattern of the Ashram School was most suitable for tribal areas. We extended this pattern by opening two more Ashram Schools at Talasari and Waki. The Ashram Schools have played an important role in laying the foundation of primary education among tribals.

Due to lack of educational facilities the percentage of literacy among the tribals was very low. We, therefore, started adult literacy classes. Since the tribal men and women went to work during the day time, classes for them were conducted at night in their hamlets. In addition, literacy classes were held in the afternoon for ladies who were working in the fruit orchards. We called them "Garden Schools". We received cooperation from the fruit growers of the area. The tribal men and women who attended literacy classes became conscious about health and sanitation. They adopted improved practices in agriculture and increased their food production. Some of the women also opened up national savings accounts. We found that socio-economic awareness could be generated among tribals through literacy programmes.

We established the Krishi Vigyan Kendra to upgrade the skills of tribal farmers. The theme of the training provided was "learning by doing". We found that the Krishi Vigyan Kendra played an important role in the transfer of technology among tribal farmers. The tribals in the area were facing problems of hunger and malnutrition. We, therefore, gave first priority on boosting their rice production. In the fifties and mid sixties we introduced improved varieties of rice while in late sixties high yielding varieties of the same were introduced. Cooperatives were started for the supply of inputs such as seeds and fertilizers. This enabled the tribals to become self-reliant in food production.

A programme of sinking small irrigation wells was undertaken on the lands of tribals. Pumps were installed on these wells for lifting water. The Chief Minister of Maharashtra, Shri Vasantrao Naik took special interest in this project and instructed the Maharashtra State Electricity Board to create necessary infrastructure for supply of power. This enabled the tribals to grow multiple crops thereby getting gainful employment on their lands.I found that land reforms and infrastructure development has helped greatly to improve the productivity of lands of tribals. It has also helped to generate gainful employment which is the need of the day.

3

LEADERSHIP DEVELOPMENT

Work Experience Projects

One day in the year 1957 Acharya Bhise came to Kosbad and told us that we have to start one Ashram School at Kosbad. The scheme of the Ashram School was recently started in the country. This scheme originated in the State of Orissa. The Chief Minister of Orissa, Shri Nabha Krishna Chaudhari and his wife Shrimati Maltidevi Chaudhari were committed social workers. They started residential primary schools in tribal areas. Due to these schools it was possible for the tribal children from isolated hilly areas to receive education. The teachers were staying in the schools with the students. The education imparted in these schools was not simply academic but used to contain extra curricular activities to inculcate idealism in their life. The atmosphere of these schools were similar to Gandhiji's Ashram. Shri Nabha Krishna Chaudhari therefore coined the name "Ashram Schools". The Government of India accepted the pattern of Ashram schools and prepared a scheme for its implementation throughout the country.

The first Ashram School in Thane district was started at Talwada in Jawahar Taluka. Adivasi Sewa Mandal was the implementing agency. Land measuring ten acres was acquired to set up a training farm. Acharya Bhise entrusted me with the responsibility of developing this farm. I was regularly visiting Talwada and used to stay there. Shri Mahajan Guruji a saintly person was the organiser of this school. He was a bachelor and committed his life to the service of the tribals. He used to live with students and take the same food as the children. The tribal children had great respect for Mahajan Guruji. The whole atmosphere of this school was like Gandhiji's Ashram.

I was greatly impressed by the atmosphere of the Ashram school at Talwada and readily agreed to the proposal of Acharya Bhise to start an

Ashram School at Kosbad. Accordingly we started this scheme in 1958. A piece of land measuring about five acres was purchased for construction of the building and setting up a training farm.

The foundation stone of the building of Ashram school was laid by Shri Kedarnath Kulkarni who was popularly known as 'Nathji'. He was a saintly person and great scholar of spiritualism. He had written several books on this subject. His book *Vivek Ani Sadhana* (Conscience and Devotion) is well known. Shri Kishorlal Mashruwala who was colleague of Mahatma Gandhi used to regard Nathji as his *Guru* (Teacher) Nathji had a large number of disciples in Gujarat and Maharashtra. His brother, Shri T.A. Kulkarni, was Chairman of Gokhale Education Society.

The construction of the building of the Ashram School was completed within a year. This building was inaugurated by Shri C.D. Deshmukh, Chairman of University Grants Commission and former Finance Minister of India. Smt. Durgabai Deshmukh was also present. Shri Deshmukh was a friend of Acharya Bhise. He praised the selfless services of Acharya Bhise in the cause of tribal welfare. He also hoped that the students passing from the Ashram School would continue the tradition of selfless service of Acharya Bhise.

We tried to create the atmosphere of an Ashram in this school. The teachers were staying with the students which helped to foster a close bond between them. The students regarded them as their parents and caretakers. There was a morning prayer followed by ablutions. After that, the students went to the farm for their work education projects. The objective of these projects was learning skills in agriculture, animal husbandry, poultry, fisheries and village industries.The theme was 'learning by doing'.

Vegetable gardening was one of the popular projects of tribal children. They were given a plot of land where they learned to grow a variety of vegetables during different seasons. The produce from their project was utilised in their kitchen. The students got the satisfaction of consuming the vegetables they grew. The students selected another project of budding and grafting of fruit trees. They learned the skills of budding of *ber* and grafting of mangoes. The students also undertook a project of selection of seeds, for which they went round the rice fields and selected best panicles of rice. They learned the skills of preservation of seeds. Since rice is the main crop of tribals every student selected a project of growing improved varieties of rice. They were given small plots for this purpose. The students learned the skills of preparation of rice nursery, green manuring, transplanting, interculturing, manuring, plant protection, harvesting and threshing. The tribal villages are surrounded by forests and hence the students were eager to learn about growing various trees. They selected a project of preparing a nursery of fuelwood and timber trees. They also learned the skills of planting and caring for trees.

The students selected a project of rearing of calves. They learned the skills of feeding the calves with nutritive fodders. They also learned about deworming of calves. One of the popular projects among students was poultry keeping. They learned the skills of care of one day old chicks, feeding of birds, treatment of diseases and collection of eggs. There was a small pond on the farm where the students learned the skills of rearing fish. Another popular project was bee-keeping. The students learned the skills of capturing bees from hollow trunks of trees, keeping bee colonies in bee boxes, care of the boxes, extraction and preservation of honey. The students also learned skills of the village craftsmen like the carpenter and blacksmith, palm *gur* making, and preparation of straw covers. The girl students learned the skills of preparation of pickles, *papads*, processing of cereals and pulses, besides embroidery and tailoring.

The students applied these skills on their lands. Whenever they went on vacations a sample of seeds, plants, budding sticks and hatching eggs were supplied to them as nucleus material. The students helped their parents to grow high yielding varieties of rice, they grew vegetables in their backyards, they transformed local *ber* and mango trees into improved varieties by budding and grafting techniques. They grew fast growing fuelwood species of trees on their uplands. They grew nutritious fodders for feeding their calves. They started small poultry units by introducing improved breeds. The girl students made pickles from mangoes and berries.

We have observed that if the students are given opportunity of learning skills through work education projects in their school they develop creativity in their life.

Students and Social Service

We started some projects in surrounding tribal villages, to inculcate the spirit of social service among the students of the secondary school.

The tribal farmers were using local varieties of rice. These varieties were tall with the result that they lodged on the ground with heavy winds. This resulted in low yields. The new high yielding varieties like I.R-8, Jaya and Ratna were dwarf and did not lodge on the ground. In order to introduce dwarf varieties of rice, the students prepared small packets of seeds each weighing one kilogram, and distributed them, one per tribal family. The students participated with farmers in raising nursery, transplanting, interculturing and plant protection of rice crop. A community plant protection squad was organized for that purpose. The stand of the rice crop grown by this method was very good. The students organized demonstrations of these plots and invited neighbouring farmers. They explained the salient features of the new high yielding varieties to the tribal farmers who assembled for demonstrations.The farmers were greatly impressed to see the performance of these varieties and wanted to adopt these high yielding varieties on their lands during the next kharif (Rainy) season. The tribals were in great need of food and so they accepted these high yielding varieties of rice. In this way the students learned the methods of demonstrations for introduction of high yielding varieties of rice.

The tribal farmers were not getting sufficient organic manures and hence the rice crop could not get required nutrients. Besides, the water holding capacity of these soils became less. With a gap in the rains the crops withered due to lack of moisture in the soil. To overcome the shortage of organic manure we selected a green manure plant called 'glyricidia'. This plant is fast growing and yields about hundred kilograms green leaves per year. The green leaves are applied to the soil before transplanting of rice. The students collected the seeds of this plant in May and prepared a nursery on the farm of the Kosbad Institute. They planted these seedlings on the lands of every tribal farmer. The seedlings grew fast and within three years the trees started yielding green leaves. The tribal farmers now regularly use green manure for their rice crop. There was significant improvement in their rice yields. Thus the students learned introduction of green manure for rice crop.

The diet of the tribals mainly consists of rice and hill millets. They seldom consume vegetables and fruits which are important sources of vitamins and minerals. Due to lack of protective foods there is a widespread problem of malnutrition. To overcome this problem the students selected a project of 'Nutrition Gardens'. They prepared a nursery of two vegetable plants viz. Moringa and Sesbania grandiflora and three fruit trees viz Mango, Custard Apple and Amla. The seedlings of these five plants were planted in the backyard of every tribal hut. The girl students showed receipes of vegetables to tribal women. Due to nutrition gardens the tribals now get vegetables and fruits. Thus the students learned the production of nutritive foods through self help programme.

The tribal women wake up early in the morning and go to the forest areas for collection of fuelwood. This drudgery of women of carrying fuelwood on head has been going on for generations. The students of the secondary school therefore took up a programme of planting fast growing fuelwood trees on the lands of tribals. They prepared a nursery of Subabhool, Casurina, *Prosopis juliflora* and *Acacia auriculiformis.* The tribals have a piece of wastelands, this land was utilized for planting fuelwood trees. Production of fuelwood on their own land reduced the drudgery of tribal women. The students thus learned the use of wastelands for planting fuelwood trees.

The students undertook a project of planting mango trees on lands of tribals. The Alphanso, *Payari* and *Kesar* varieties of mangoes fetch good price in the market. However, the grafts of these varieties are expensive and are not within the reach of poor people. The students therefore adopted low cost technology. The seeds of local mangoes were sown on bunds of rice fields at the beginning of rains. When the seedlings were three years old they were grafted with Alphanso, *Payari* and *Kesar* varieties. This technique is known as side grafting. The budsticks required for grafting were brought from the Kosbad Institute. In this way the students transformed the local mango trees into improved varieties. The farmers received good income through sale of mango fruits. Thus the students learned low cost technologies for introduction of horticulture on tribal lands.

There are *ber* plants of local varieties on lands of tribals. The students budded them with improved varieties like *Mehroon*, *Gola* and *Umran*. The budded plants gave cash income to the tribals. The *ber* is very hardy plant and grows on poor soils under rainfed or dry conditions. The students thus learned utilization of local *ber* plants for income generation.

There is a small platform near the hut of every tribal family. The tribals use this platform for washing clothes and cleaning utensils. With a view to utilize wash water the students took up a project of planting coconut trees. Accordingly hundred seedlings were planted by them in one village. The Hexamar Agricultural Research and Development Foundation a non-government organization in Bombay gave financial support for purchase of coconut saplings. We named this project 'One family one coconut'. The trees started bearing coconuts after 4/5 years. Thus the students learned to grow coconuts by recycling of waste water.

The girl students undertook a project for beautification of tribal huts. The serial huts do not have windows and so there is no light and aeration inside. The students provided simple bamboo windows and glasses for this purpose. They also installed smokeless *chulhas* (stoves). This has saved the fuelwood.

The students planted drought resistant flowering shrubs like bouganvillas. The girl students thus learned beautification of homes of poor people. The students of secondary school both boys and girls learned socio-economic conditions of tribals through these projects. These projects have helped to inculcate spirit of social service among the students.

Leadership Development Through Higher Education

Soon after Independence, a number of schemes for tribal development were started in India. However, the manpower required for implementation of these schemes was not available from amongst the tribals. The main reason was lack of education. Primary schools were opened for the first generation learners after Independence. However, facilities for secondary education were meagre.

We started a secondary school at Kosbad Hill in 1965. This school was inaugurated by Acharya Bhise a great sarvodaya worker who devoted his life for service of tribals. My wife, Meena, accepted the responsibility of running this school. She had a background of tribal welfare work as her father Shri Khanderao Save worked for the welfare of tribals for over two decades. Shri Save wrote a book on the Warli tribe (called The Warlis).

Shri Baboo Bujad who joined as a teacher in this school took great interest in the education of tribal students. The students in the school were proud of Shri Bujad who hailed from tribal community. He belonged to Chembave village. He used to walk 6 kilometers everyday for receiving secondary education in school at Bordi. He passed his Secondary School Certificate Examination. One day he came and told us that he wanted to pursue higher studies but could not since his economic condition was very poor, his parents being agricultural labourers. So we advised him to take agricultural education and also assured him financial support. Accordingly he went to Dapoli in Ratnagiri district to seek admission at the Agricultural college. We provided him support for his travelling, books, hostel expenses etc. Later on he received a scholarship meant for Scheduled Tribe students. He successfully completed four year course and passed Bachelor of Science (Agriculture) degree of Agricultural University. He was the first agricultural graduate from tribal community in Thane district. We requested him to work as teacher in the secondary school at Kosbad Hill and he readily agreed to our proposal. Shri Bujad was deputed for teachers training and obtained his B.Ed degree. He also received M.Ed degree from Bombay University. He became the Head Master of the secondary school in 1983.

The teachers in the secondary school tried to convince the students that they should study hard and go for higher education and utilise their education for welfare of thier community. The teachers gave great encouragement to these first generation of learners.

We found that the tribal students were very sincere in their studies. Some of the students scored good marks in their secondary school certificate examination. Two students from this school viz. Shri Bhoye and Shri Kalangada successfully completed their medical education and obtained M.B.B.S. degrees from the Bombay University. Both of them returned to their areas and are serving the tribal people. Shri Gangoda, Shri Godale and Shri Bhoye have obtained Bachelor of Science (Agriculture) degree and are doing agricultural development work among the tribals. Besides about seventy girls have obtained diplomas in education and are serving as teachers in Ashram School in tribal areas. Some girls have completed training in nursing and are working in hospitals. About hundred students have completed two years course at the Agricultural School and are working as village level and extension workers in the tribal areas. Some of the students who completed agricultural training courses have returned to their lands. They have adopted modern practices in agriculture and have improved their agricultural productivity. Their farms are proving to be demonstration centres to the neighbouring tribal farmers.

Shri Arjun Uradya a tribal student after completion of his secondary education at Bordi passed M.Sc. (Agriculture) from Konkan Agricultural University with plaudits. He has been working in Krishi Vigyan Kendra and training the tribal farmers in horticulture. He has motivated a large number of tribal farmers to plant fruit trees on their farms. Shri Uradya has undergone leadership training at the International Institute of Rural Reconstruction in Philippines. He is creating constructive leadership among tribal youths.

Shri Mahadeo Gopal Kadu and Shri Pujari Ahire two tribal students who completed a two years course at the Kosbad Institute were elected to Maharashtra Legislative Assembly. They made significant contribution in the development of tribals.

We have observed that constructive leadership is developed if the tribal students are given opportunity to receive higher education. After completion of their higher education they are willing to return to their villages to serve their communities.

Talasari Centre

Talasari is a small village inhabited by tribals. During the pre independence days, it was isolated due to lack of communication facilities. During the British rule, landlords and forest contractors committed atrocities on tribals and exploited them. The tribals therefore revolted during the mid forties in what is known as the 'Warli revolt'. The Congress ministry during pre-independence period had passed tenancy laws to protect the tribal tenants from eviction by landlords. Forest Labour Cooperative Societies were started to protect the tribals from atrocities and exploitation by forest contractors. Hostels and schools were started for tribal children for spread of education. The Adivasi Seva Mandal under guidance of Acharya Bhise started an Ashram at Talasari in the early forties. The Ashram had a band of devoted workers who spent their lives for service of tribals. Acharya Bhise had therefore great affinity for the Talasari Centre.

After the formation of the Maharashtra State in 1960, twenty seven villages from Umbergaon taluka were carved out and the new Talasari taluka was formed. All these villages were predominantly inhabited by tribals. There was no secondary school in the entire taluka and so Acharya Bhise started a secondary school at Talasari. Shri Nathubhau Arekar was appointed as Head Master of this school. Shri Arekar was educated at Bordi under the guidance of Acharya Bhise and so he had developed the spirit of social service.There were no buildings for the school and so the classes were held in the buildings of the Ashram. The responsibility of conducting this school was entrusted to Gokhale Education Society's Bordi Centre.

After completing my studies in the U.S. I returned to India in 1964. I was going to Kosbad Hill from my home town of Bordi for about two months. One evening Acharya Bhise came to my house alongwith Shri S.V. Apte who was principal of the secondary school at Bordi. Acharya desired that the responsibility of conducting the secondary school of Talasari be accepted by the Kosbad Centre. He further proposed to start an Ashram School and agricultural farm at Talasari. After consultation with Shri Harischandra Patil, Director of the Kosbad Institute we accepted the responsibility of conducting the Talasari Centre.

The Government of India prepared a scheme of Post Basic Ashram Schools to give primary and secondary education to tribal boys and girls. Acharya Bhise requested the Maharashtra Government to sanction one such school at Talasari as there was no secondary school in the whole of Talasari taluka. Shri Madhukarao Chaudhari was Education Minister of Maharashtra State at that time. Acharya Bhise used to meet him several times to get this scheme sanctioned. Many a times I used to accompany him. Due to the untiring efforts of Acharya Bhise, the Government of Maharashtra sanctioned the first Post Basic Education Ashram School in Maharashtra at Talasari.

We invited Shri Madhukarao Chaudhari to lay the foundation stone of this Post Basic Ashram School. Shri Chaudhari also held the forest portfolio. We requested him to sanction 40 acres of land for starting a training and demonstration farm. My father who was a colleague of Shri Chaudhari in the Maharashtra Cabinet also convinced him about the need of land for the school. Shri Madhukarao Chaudhari immediately sanctioned this land and instructed the divisional forest officer Shri Khot to hand it over. Accordingly all the documents of the land were given to us at Umbergaon Road Railway Station before Shri Chaudhari left for Bombay. We named this Institution the 'Thakkar Bappa Vidyalaya'. Late Shri Thakkar Bappa had devoted his life for upliftment of tribals and it was appropriate that this Institute was named after him.

The land needed development, besides fencing, irrigation wells, pumping sets and planting material for the demonstration and training farm A capital expenditure of rupees one lakh was required to implement this project, which was beyond our financial resources. We requested Mr. Neil O'Sullivan, Director of Community Aid (Australia) to visit Talasari whenever he visited India. During his visit we took him to Talasari and showed the site of the proposed farm. He kept a tape recorder before me and told me to state the objectives, programmes, and financial implications of the project. Mr. Sullivan carried this tape to Australia. After a few days we received a letter from Community Aid Abroad (Australia) in which they informed that since this project was meant for training of students from weaker sections they were happy to sanction a sum of one lakh rupees for its implementation. We developed this farm within one year. The farm was useful in imparting practical training to the tribal students. Besides, the seeds, plants and grafts produced on the farm were supplied to the students to put training into practice on their lands. The support of Community Aid Abroad (Australia) was crucial in setting up this farm in time.

Major Ramchandra Salvi, Secretary to Government of Maharashtra, Education and Sports Department once visited the Talasari Centre. After seeing the tribal welfare work at the Centre he considered it an appropriate place to inculcate the spirit of social service among the University students. He therefore initiated a scheme for setting up a youth centre at Talasari. We constructed a hostel building and a community hall. The students of Bombay University regularly came for camps. They participated in various tribal welfare activities. Many students continued their activities even after their graduation. We found that if university students are exposed to rural development activities they empathize with rural issues and inculcate a spirit of social service.

The Post Basic Ashram School gave the tribal students from Talasari

Taluka an opportunity to appear for Secondary School Certificate examination. It was the first generation of learners. Some of the students passed the S.S.C. examination, entered college and obtained University degrees. They were not tempted by cities and returned to their tribal roots to serve their communities. Shri Chintaman Vanaga obtained his B.Com. and L.L.B. degrees, but returned to Jawahar to serve the tribal people. Shri Mahesh Kondhari obtained his degree in Veterinary Science and returned to Krishi Vigyan Kendra at Kosbad Hill to train the tribal youths in animal husbandry and poultry. His brother Dr. Kondhari also returned after obtaining M.B.B.S. degree to start medical service among the tribal people.

Acharya Bhise had a dream that the students of Post Basic Ashram school would utilize their knowledge for the upliftment of the tribal community. His dream has been realised by students returning to their roots to do welfare work among the tribals.

Youth Clubs

While I was studying in the U.S.A. I got the opportunity to study the 4-H Clubs. The word 4-H comprises head, heart, hands and health. Rural youths should learn good things in life by actually doing the work is the motto of these clubs. The youths between the age of ten to twenty years are enrolled as members of these clubs. The members have to take following pledge :

> " I take a pledge that I will use my head for good cause, my heart for devotion, my hands for greater service and health for living a good life for my club, my community and my country."

To learn anything in life by actually practising is the philosophy of these clubs. The young boys and girls select different projects and implement them in their homes, farms or in surrounding areas. Rearing of calves, sheep, pigs, dogs and horses, poultry-keeping, bee-keeping, preparing fruit products, vegetable gardening, plant protection, are some of the projects selected by them. Besides they entertain the community through dances, dramas and cultural programmes.

On my return to India I started such youth clubs in five villages, Bordi, Gholvad, Chikhale, Narpad and Kosbad. We first organized rallies of youths comprising both boys and girls. Most of the students were from secondary and higher secondary schools. We also invited the teachers to these rallies. There was a convention of teachers at Mahalakshmi where I narrated the concept of the 4-H Clubs in U.S.A. This was helpful for creating an atmosphere for establishment of Youth Clubs in our region.

We selected simple projects which were within the reach of every member. The first project was of " vegetable garden". The seedlings of 5 vegetables viz. egg plant, tomato, onion, garlic and chilly were prepared in the nursery of the Kosbad Institute. We organized a meeting of the Youth Club members at the Institute and gave them demonstrations about planting and care of these five vegetables. Every member had a small backyard around their house. They prepared the beds, applied manure and planted the seedlings. We also showed them simple methods of plant protection: BHC powder was tied in a muslin cloth and dusted on the leaves of the seedlings. This controlled leaf eating caterpillars and other insects. We gave them handouts in which cultivation practices of these five vegetables were written in simple language. The youth club members were observing the height of the plant, number of leaves, number of flowers and fruits. We supplied them with a low cost balance. They recorded observations about the number of fruits and their weight from each plant. This has inculcated habit of observation of plants. They were discussing with each other about the performance of their plants. There was a close interaction among them.

We arranged an exhibition of the vegetable produce at Gholvad. Every member participated in this exhibition by bringing a sample of vegetables which they grew in their backyards. The teachers of primary and secondary schools cooperated in this exhibition. We invited the parents of the youth club members to watch the exhibition. We arranged a seminar in this workshop. Each member narrated his or her experience about growing these five vegetables. The seminar was very interesting as every youth got opportunity for the first time to express their experiences. That has created a self confidence among them. We distributed prizes for the best samples. The members utilised vegetables produce for their home consumption. Some of them earned cash income through sale of extra vegetables. They utilised this amount for study tours, buying books, and purchase of small saving certificates. Some members purchased mango and chikoo grafts to planting in their backyard. The members were very proud as they could generate cash income through their labour.

The youth club members started other projects like poultry keeping, tailoring, embroidery, bamboo and cane work, pickle and *papad* making. We gave them support to start a small library in five villages where books on agricultural subjets were kept. Besides, magazines on agriculture and rural development were also kept in this library.

The members were celebrating festivals like *Nag Panchami* (worshipping cobra). They explained to the villagers about the importance of reptiles for their role in control of rats and insects. The girl members celebrated *Vat Purnima* (worshipping Banyan Tree). They explained to the village women about the importance of preserving banyan trees. The members were also celebrating a festival of *Pola* (worshipping of Bullocks), the day when they fed sweets to the bullocks. The members of Youth Club explained the role of plants and animals in preserving the environmental balance.

Since these projects were based on the principle of 'learning by doing', the spirit of dignity of labour was created amongst the youth. They learned the principle of self reliance. They also felt that their projects were useful not only to themselves but to the community and the nation. These projects had created constructive leadership among the youths.

National Service Scheme

The girl students from Sophia College in Bombay came to the Kosbad Institute for their Programme of National Service Scheme. We used to guide the students for undertaking social service projects among the tribals. The students had adopted a hamlet called Dalvipada for that purpose. Ms. Ladia was the team leader of this group. She was an enthusiastic lady who used to motivate the students to participate actively with tribal women in their agricultural activities. The girl students helped the tribal women in preparing raised beds, sowing of seeds, transplanting and harvest of rice crops.

The students prepared the seedlings of drumsticks *(moringa)*, *Agati*, Custard Apple and *Amla* in polythene bags. They helped the tribal women plant the seedlings in their backyards. In this way they helped them to establish Nutrition Gardens.

The tribal women from Dalvipada had to walk long distance to fetch water from a nearby stream, which was however unclean. The students therefore, decided to construct a drinking water well. For that purpose they raised funds through cultural programmes in their college. The tribal women selected a site near the hamlet for digging the well. The girl students offered their voluntary labour in digging this well together with the tribal women. Seeing the hard work of the girl students the tribal men from Dalvipada hamlet also participated in digging the well. The water was struck at the depth of 12 feet. A masonry construction was done on this well. A permanent drinking water well was constructed for the Dalvipada hamlet through the efforts of the girl students from Sophia College. The inauguration of this well was done by the Principal of Sophia College, Mother Bregenza. One of the tribal women spoke at this inaugural function. She said that well had reduced the drudgery of carrying water from long distance. Besides the people in the hamlet could now get clean drinking water which would improve their health. She complemented to the girl students of the Sophia College for their noble gesture. It is a novel example of how the student community could be helpful in the community development programme.

The students of the Sydneham College of Commerce and Economics, Bombay also started coming to Kosbad Institute for the National Service Scheme and adopted Waki and surrounding villages for this purpose. Their residential arrangements had been made in Ashram school at Waki. They too undertook a programme of constructing a community well and raised funds for installation of a pumping set on this well. Fifteen tribal farmers got the benefit of this community well. These farmers grew only one crop of rice and did not have any gainful employment on their land for the rest of the eight months. With the construction of the community well they started growing vegetables after the harvest of rice. Growing of vegetables generated cash income thereby improving their purchasing power, besides receiving gainful employment for eight months on their lands. Due to the inclusion of vegetables in their diet the problem of malnutrition could also be overcome. The students thus learned the importance of irrigation in generation of employment.

Dr. Bal the Principal and Prof. Abhayanker, Vice Principal of the college came regularly with their students, often accompanied by other professors. The staff and students of the Sydneham College decided to construct a permanent hostel for their NSS work at Waki. They collected donations for that purpose. The hostel building was constructed at Waki in the campus of the Ashram School. The inauguration of this building was done by Shri Vijay Merchant a well known cricketer of India. Shri Merchant was an alumni of the Sydneham College and he was very proud that students of the college were doing social service among the very poor sections of the society like tribals.

The students organized a health camp in which renowned doctors from Bombay were invited. Large number of tribals attended this camp. The health camp became a regular activity under NSS scheme. The doctors attending the camp referred some of the serious cases for further treatment in the hospitals in Bombay. The students offered every possible help to the tribal patients for hospitalization in Bombay. This was an important humanitarian work carried out by the students for the tribal people.

National Service Scheme is a useful means of giving an opportunity to the students to serve the poor and the downtrodden of our country. It is through such service that a spirit of social service is inculcated amongst the youth.

4

HEALTH AND NUTRITION

Drinking Water Wells

Bhonarpada is a small tribal hamlet having 50 to 60 huts in Kainad village. The name of the hamlet was derived from the surname Bhonar of the Warli tribe. A hamlet is called *Pada* in tribal dialect. We used to visit Bhonarpada with students of the Kosbad Institute to survey agriculture in this hamlet. We wanted to introduce improved varieties of rice among the farmers of this hamlet, so that their yields could be improved. We went from house to house and convinced the farmers about the importance of using improved seeds. Shri Chamar Bhonar was leader of this hamlet. He was a straight forward person with very pleasant nature. He explained to the tribal farmers from his hamlet that the Kosbad Institute was arranging to supply improved seeds of rice which they should utilize and improve their rice yields. All the farmers had great faith in Shri Bhonar and they agreed to adopt the new improved varieties of rice. We were planning to arrange demonstrations during the coming monsoon season.

While returning from the village we came across a small ditch from where a tribal girl was fetching water. The girl was using an earthern pot. On seeing this a calf came running and started drinking water. It passed urine which washed towards the ditch. We asked Shri Chamar Bhonar whether they were utilizing the same water for drinking purposes, and his reply was 'yes'. He told us that they don't have a *pucca* well in the village and so they have to drink this polluted water.

We felt very sorry to see that people within one mile radius of our Institute have no facility of clean drinking water. However, simply by expressing our emotion was of no use. Concrete action was necessary. We immediately took a decision to construct a drinking water well for this hamlet. We requested Shri Chamar Bhonar to spare a piece of land for construction of a well. Fortunately a small piece of land adjoining to this ditch belonged to Shri Bhonar. He agreed to give this land for construction of the well. Next morning, the students and staff of our Institute came to the village alongwith crowbars, pick-axes and iron baskets. Shri Chamar Bhonar with all the tribal men, women and children was there to welcome us. As a traditional ceremony we did *Bhoomi Poojan* (worshipping the soil) and broke a coconut and the students of the Kosbad Institute recited *shlokas* (hymns) from *Ashram Bhajanavali.*

When we started digging, the tribal people assembled there also joined us in *Shramdan* (voluntary labour). Within a week we had dug upto about 10 feet deep. Traces of wet mud had started appearing. Shri Chamar Bhonar said that this was an indication of water. Water was finally tapped at the depth of 12 feet. We brought empty tins and ropes to take out water as pump sets were not available at that time. The tribal people from the hamlet used to wake up early in the morning to draw water from the well. This had helped to dig the well during the morning hours.

Shri Jasjit Singh, the Collector of the District came to visit the Agricultural Institute. When he heard that the students and staff of the Kosbad Institute were constructing a drinking water well at Bhonar Pada he came to the site. Shri Khanderao Save, Deputy Collector of Dahanu accompanied him. Shri Save told Shri Singh that most of the tribal villages do not have drinking water wells and hence the people have to drink polluted water from ditches and streams. Various diseases and ailments were caused due to lack of clean drinking water. Shri Singh immediately responded by promising to make available discretionary grants (at his disposal) for the construction of drinking water wells. Shri Singh congratulated the students and the staff and assured us that he would send necessary grants for masonary construction of this well. He further assured us if we construct such wells in other hamlets he would be happy to give financial support for masonary construction.

When the well was about 15 feet deep and had enough water, we started masonary construction of the well. The stones were available in the nearby hill. The stone cutters who are locally known as *Vadaris* undertook the work of preparing stones for construction. The farmers from Bhonarpada brought the sands in their bullockcarts from Chikhale, a nearby coastal village. The cement was purchased from Dahanu Town. The masons from Chikhale village undertook the construction work. The well, 12 feet deep and 12 feet in diameter, was completed within a fortnight.

Shri Appa Saheb Patwardhan, a great constructive worker and follower of Mahatma Gandhi came to the Kosbad Institute. The well was inaugurated at his hands. It was a coincidence that a saintly person like Shri Patwardhan was available on such an auspicious occasion. All the tribal men, women and children from Bhonarpada were present for this ceremony. They were very happy as clean drinking water would be available to them.

In tribal parlance, a well is called a *Bao*. Tribals from the surrounding hamlets started coming to see this *Bao* in Bhonarpada hamlet. They also felt that they should have such a *Bao* in their hamlet. Two tribal farmers, Shri Nausha Hadel and Shri Devji Tandel from Lilakpada hamlet, came and told us that they did not have a drinking water well in their hamlet. They requested our help in constructing a well. We agreed to their proposal and started digging the well in their hamlet. This well also struck good water. We completed the masonary work and well was inaugurated by Shri Nana Saheb Tilak, a great social worker from Nagotone, Raigad District.

Seeing the well at Lilakpada Shri Radaka Bendu and Shri Damo Daware, two tribals from Zarli hamlet approached us for construction of a drinking water well in their hamlet. We agreed to their proposal. When we went to dig the well a number of tribal women assembled there. They told us that there was one well in the village which belonged to a landlord. However, when they went to fetch water from this well, the watchman threw away their earthern pots and did not allow them to draw water. They were therefore in great need of a community drinking water well, and very happy that such well was being constructed in their hamlet. All the tribal women from Zarli village joined us in digging the well. There was good supply of water in this well. We completed the well within two months. All the tribal people were very happy as the well belonged to the community, and now they could fetch water with self respect.

We have observed that due to the construction of drinking water wells there was a significant improvement in the health of the people. The tribals who were deprived of clean drinking water for generations could now get clean drinking water from their own community wells.

Gopuri Latrines

The Public Works Department of Government of Bombay undertook construction of buildings of the Agricultural School in 1950. At that time Shri Mahida was the Superintending Engineer while Shri Bakeri was the Executive Engineer. Acharya Bhise told them we require latrines which could process human waste so that good manure could be prepared out of it. Both these engineers told him they had the schematic patterns of only two types of latrines. The basket type in which night soil had to be carried by scavangers and the type using septic tanks. Acharya Bhise told them it would be better if they construct the Gopuri type latrines which were designed by Shri Appa Saheb Patwardhan at Kankavali in Ratnagiri District. However, the engineers expressed their inability to construct such latrines as they were not approved patterns of the Public Works Department. Acharya Bhise then approached Shri Vaikunth Lal Mehta who was the Finance Minister of the Government of Bombay. The design of these Gopuri latrines were secured from the Kora Gram Udyog Kendra at Borivali where such latrines were constructed. The Government of Bombay gave approval for the construction of the Gopuri latrines and for the first time the Public Works Department constructed Gopuri latrines at Kosbad Hill.

Shri Appa Saheb Patwardhan once approached Mahatma Gandhi and asked him what he should do in his life. Gandhiji told him that he should devote his life for sanitation. Accordingly, he established an Ashram in a very remote town called Kankavali in Ratnagiri District. He named his Ashram Gopuri. He took up the programme of preparing night soil manure by recycling of human waste. Every morning he collected night soil from the latrines of the town and brought it to his ashram at Gopuri. He prepared night soil compost which was found very useful to rice crops, vegetables and fruit trees. Appa Saheb then designed a latrine which he named as Gopuri Latrine. There are two tanks of 1 x 1 x 1 metre size. These tanks are to be used alternatively. When the first tank is filled it is to be covered by soil and the other tank is to be used. The night soil in the first tank is decomposed in five to six months and good manure is obtained from it. These latrines are very simple to operate. Besides, very little water is required as there is no flush system. Both the tanks have wooden lids and hence they are closed after the use. Due to these lids there is no menace of flies. Two ventilation pipes are also provided for emission of gases. These latrines could convert human waste, obtained from a five member family, into two tonnes of manure every year.

The Gopuri latrines constructed at the Kosbad Institute became a demonstration to the farmers who were visiting the Institute. They constructed such latrines in their homes. At present due to lack of facilities of latrines in the villages, people squat in the open, creating unhealthy conditions. If cheap type latrines like the Gopuri are constructed, it would help to solve the problem of sanitation and also create wealth out of waste.

Starting of Medical Facilities

This is an incidence of 1953. Shri Chamar Bhonar and two tribals from Bhonarpada hamlet came to me one early morning, at 6'O Clock, and told me that they require the bullock cart of our Institute. It was the rainy season. During this season, the tribals take out the wheels of their bullock carts and tie them on their mud walls as the bullocks are used for agricultural operations like ploughing, puddling etc. The roads are so muddy that hardly any tribal uses his bullock cart. Shri Chamar Bhonar told me that one woman from their hamlet had difficulty in delivery and had to be moved to a maternity home at Gholvad village. We immediately gave them the bullock cart and sent one of our staff to help.

The doctors at the Gholvad maternity home informed that because of complication, surgery would have to be performed. The tribal nurse in the village (locally called *suin*) could not perform the delivery and had used crude methods to pull the child. As a result, the child was dead and septic had formed in the body of the woman. The doctors advised to take the patient to the Mission Hospital at Dahanu Road. As there was no *pucca* road for going to Dahanu, and there were no other means of transport, the only option was the railway. We therefore arranged to bring her to Gholvad railway station and took her by train to Dahanu. The Station Master at Gholvad was very helpful. He instructed the driver and the guard to detain the train till the patient was accommodated.

The woman reached the Mission Hospital at Dahanu by 12 noon. She was in coma. Dr. Secter of the American Mission Hospital immediately admitted the case and performed the necessary operation. He tried his best to save her life but did not succeed. The dead body of the woman was brought from Dahanu Road to Bhonarpada by bullock cart in the evening. We all attended the funeral at night.

After this incidence we requested Dr. Secter to visit Kosbad Hill once in a week so that medical check up of such cases could be carried out. He readily responded to our request. There was no pucca road from Dahanu Road to Kosbad. However, he used to come by Jeep via Chikhle and Gholvad by a *kuccha* road. The road was not in good condition, but he tolerated the inconvenience. We gave one room from the Kosbad Institute for his medical centre. Our staff and students used to go around the villages and advise the tribals to come for medical check-up. The patients included pregnant and nursing mothers, infants, children and old people. Dr. Secter used to examine them and give them medicines. This medical centre proved to be a boon to the surrounding areas.

Unfortunately, Dr. Secter's wife died due to sudden illness and he had to return to the United States. Thereafter Dr. Wampler and Dr. Waybright came from United States. They too visited Kosbad Hill regularly to give medical help. They used to take serious cases in their jeep to the Mission Hospital at Dahanu Road. The tribals of the area developed great faith in medical treatment. Hitherto they were at the mercy of a local tribal "faith-healer" whom they called *Bhagat.* In this way, the first generation of tribals started taking medicines. Thereafter, we started a regular dispensary. Dr. Dinanath Churi, a local doctor from Bordi who was a committed social worker took the responsibility of running the dispensary. As there was no *pucca* road for coming to Kosbad Hill from Bordi he used to come in a horse tonga. He laid the foundation of the dispensary at Kosbad Hill, which greatly helped the tribals to get medical service.

The founder of the Mission hospital of American Brethren Mission at Dahanu was Dr. Barbara Neki. She laid the tradition of selfless service in this hospital. Barbara was a very intelligent student during her school days. After she passed her M.D. Examination from the United States, she decided to use her medical knowledge for the poor and accordingly, came to Dahanu Road in February 1915. At that time there were no medical facilities in the area and there were a number of deaths due to lack of treatment, especially of women during delivery. Dr. Neki toured the surrounding areas and created confidence among the local tribal people. She learned Marathi and Gujarati. There were no buildings, so she started the hospital in a small hut. In this effort a local social worker Smt. Shirine Bai Horomasji Irani extended her very valuable help. Dr. Neki saved the lives of hundreds of women especially during the time of delivery. She felt the need of a building for her hospital as the number of patients had increased. She requested her Mission in the United States to send a donation for the hospital building. The money was not available with the Mission. However, taking into consideration the need of the hospital the Mission appealed to the educational institutions to collect donations for the hospital. The students and staff of the educational institutions collected 2000 dollars by selling vegetables and fruits in the community. This amount was sent to Dr. Neki for building of the Mission hospital at Dahanu Road. In this way the building of the Mission Hospital of Dahanu Road was constructed through the efforts of students of the United States. Dr. Neki served for 37 years at this hospital. In 1952 she retired and went back to the United States and stayed in a small village called North Manchester in the State of Indiana.

I went to the United States in 1962 and wrote her a letter. She requested me to visit her place. She also arranged for my transport from East Lansing, Michigan to North Manchester, Indiana through one of her missionary friend. Accordingly, I visited her in April 1963 and stayed with her for three days. She was leading a very simple life. She recounted her memories of the Dahanu Mission hospital. Later on she went to California and expired there. This is the amazing account of a committed doctor who sacrificed all material comforts and came all the way from the United States to India and spent 37 years in the service of the poor and the downtrodden. This is a truly touching example of selfless service, or *Nishkam Karamyog*, in the parlance of Hindu mythology.

Sanitation Camps

Shri Krishnadas Shah a social worker from village Chinchani in Thane District had devoted his life to the cause of *Safai* (Sanitation). Once he went to Mahatma Gandhi to seek his guidance. Gandhiji was camping at Sodpur Ashram near Calcutta. Young Krishnadas told Gandhiji that he wanted to remain a bachelor and do some constructive work. Gandhiji suggested that he could take up *Safai* as constructive work. Shri Krishnadas Shah agreed to Gandhiji's suggestion and asked him when he should make a beginning. There was a broom in the corner of the hall. Gandhiji told him to take the broom and clean the hall. This would be the beginning of his *Safai* work. Since that time Krishnadas Shah devoted his life to *Safai* work.

We were aware of the devotion of Shri Krishnadas Shah to his mission of *Safai* and wanted our students to seek his guidance. We organized a *Safai Shibir* (Sanitation camp) at Bhonarpada village. Shri Krishnadas Shah agreed to attend. The Balsevika trainees from Vikaswadi Institution of Shrimati Tarabai Modak and Anutai Wagh also participated in this camp. All the participants were staying at the village with the tribals. Shri Krishnadas Shah also stayed in a small tribal hut. This created confidence among the tribals about the sincerity of the participants.

The participants used to get up early in the morning at 5 a.m. and clean the surroundings. At 7 a.m. there was a *prabhat pheri* (Morning Procession). That created awakening in the village. The tribal children, men and women also participated in *prabhat pheri*. The participants then had their breakfast which consisted of sprouted mungbeans, groundnut and jaggery. At 11 a.m. all the participants bathed at a nearby well. Community lunch was arranged at 12 noon, followed by a rest period of about an hour.

Shri Krishnadas Shah took a class on *Safai* at 2 p.m. He had developed a sense of *Safai* and in each programme of *Safai* he developed appropriate technology. The participants helped tribals to prepare a soakpits around the huts. All the wash water was diverted to this soakpit. To utilize this water, seedlings of papaya and vegetable plantains were planted near the pit. The tribals who used to defecate in the open, had now prepared trench latrines in their backyards. The tribal women benefited greatly, from such latrines. They told us that they had to go for toilet either early in the morning or late evening when it was dark. During the day, they could not go which caused number of stomach disorders. The participants prepared trench latrines and urinals for children. This helped inculcate the habit of using trench latrines from childhood. Shri Krishnadas Shah was of the view that local material like bambooes, date palm leaves should be used for construction of latrines so that they would be within the reach of poor people. There were several heaps of trash and dried leaves all over the hamlet. The participants helped the villagers to prepare compost pits. They also demonstrated how to prepare compost. The tribals used the compost in their rice fields.

The women *balsevika* trainees participated in beautification of tribal huts. They plastered the walls of the huts with red soil and cow dung. The tribal women then created paintings on mud walls. These are known as 'Warli paintings' which is a native art. The walls of the huts were prepared from a local plant called *Karvi* and bambooes. Windows were not provided in these walls and so there was no light in the hut. This caused great inconvenience to women while cooking. The participants fixed small glasses in the walls. This allowed light to enter the house. The participants also prepared simple windows from bambooes. This facilitated air circulation in the hut. Bougainvillea were planted in front of huts. This flowering shurb was very hardy and suitable for hilly and dry areas. The women participants helped the tribal women in preparing improved *Chulhas* (stoves). Tribal women had to walk long distances to fetch fuelwood. There was considerable saving of fuelwood due to improved stoves which reduced the drudgery of women. Being smokeless they were also healthier.

There were tamarind trees in Bhonarpada village. The roots of these trees were exposed due to erosion of soil. Circular platforms, locally called *par*, were prepared by arranging stones round the tree and filling with soil. Shri Krishnadas Shah used to take classes on these platforms. The children used to play on these platforms. The tribals and women used these platforms for rest in the afternoons.

Community prayers were held in the evening which were also attended by local tribals and children. The participants narrated their experiences and activities of the day. The participants were also recording their experiences, everyday in their diary before going to bed. These experiences made permanent impression on the minds of the participants about the importance of village sanitation. Within a week the Bhonarpada village had a different look. Cleanliness was everywhere. Shri Chamar

Bhonar, a respected tribal leader, from the village told Shri Krishnadas Shah that he had transformed the village. He promised him to continue the work of *safai*.

Shri Krishnadas Shah moved from village to village to organize *Safai Shibirs*. Shri Krishnadas Shah himself used to take a broom in his hand and start cleaning human excreta to prepare compost manure. He used to say that if all the waste was scientifically recycled, it could be transformed into wealth. Shri Krishnadas Shah passed away while doing *safai* work at Kolhapur in Maharashtra. He was a real missionary to the cause of rural sanitation.

Nutrition Gardens

The tribals seldom consume vegetables, fruits, milk or meat and as such their intake of protective foods enriched with vitamins and minerals is negligible. This has led to widespread malnutrition. The problem is more serious among pregnant women, nursing mothers and children. Due to malnutrition, diseases and ailments prevail among the children and adults. This was one of the reasons why children did not attend school. The adults too were unable to put in the hard labour required for agricultural operations which resulted in poor yields of crops.

With a view to overcoming the problem of malnutrition a self help programme of setting up of "Nutrition Gardens" was undertaken. The programme consisted of training the tribals, both men and women in growing vegetables and fruit trees under rainfed conditions. The tribal areas receive about 2000 mm of rainfall between the months of June to September. The remaining eight months are completely dry. It was therefore necessary to select those trees which have deep root system and which extract moisture from deeper layers of the soils during dry months. The following trees were found to be suitable both in nutrition as well as drought tolerance :

Vegetable trees

1. Drumstick *(Moringa oleifera)*
2. Agasta *(Sesbania grandiflora)*

Fruit trees

1. Mango *(Mongifera indica)*
2. *Seetaphal (Annona squamosa)*
3. Jack Fruit *(Artocarpus heterophyllus)*
4. Bor *(Zyziphus mauritiana Lamk)*

The produce from these trees is rich in minerals and vitamins.

Training formed an important component of the project. It was imparted on the principle of 'learning by doing'. The training was organized in the villages as the tribal women were unable to come out of their villages due to their responsibilities at home. Training covered the following topics :

1. Common ailments caused by malnutrition.
2. Importance of vegetables and fruits in diet.
3. Growing drought-resistant vegetable trees and fruit trees.
4. Plant propagation-grafting and budding.
5. Simple recipes of fruits and vegetables.
6. Low cost methods of preservation of fruits and vegetables.

Soon after the training, nucleus material like seeds, plants and budsticks were supplied to the trainees. This helped the tribals to put the training acquired, into practice. The nucleus material was further used for multiplication.

Five villages viz., Bordi, Gholvad, Borigaon, Zai and Aswali were selected for the project of 'Nutrition Gardens'. These villages comprise of nearly 12,000 tribals. The Sarpanch and Members of Village Panchayat were actively involved in training and implementation. Similarly, members of the village co-operatives were also invited to participate in the programme. Tribal farmers possessing irrigation facilities such as small wells were involved as demonstration farmers. Their farms were utilized for imparting training. These farmers were motivated to start nurseries for supply of nucleus material in the form of seeds, plants and bud-sticks. In this way 'People's Nurseries' were started under this programme.

Low cost technologies were developed for introducing in the Nutrition Gardens. For instance at the beginning of the rainy season the seeds of local mangoes were sown in backyards. When the seedlings attained finger thickness, they were grafted with improved varieties like Alphanso, *Kesar, Langra*, etc. In this way local mango trees were transformed into improved varieties. Shri Bhikoo Dubla from Jalwai village had transformed about 25 local mango trees into Alphanso and *Langra* varieties. Besides getting fruit for his own consumption, he also sold fruit worth 5 to 6 thousand rupees every year. Mango thus has proved to be a tree of nutrition and income generation.

Ber was yet another plant which was utilised for nutrition supplementation and income generation. Local *ber* trees which existed everywhere were transformed into table varieties by budding techniques.

Custard apple was the other fruit tree which was suitable for nutrition as well as income generation. This tree is non-browsable and hence could be grown where there was menace of stray animals.

Drumstick and Agasta were fast growing vegetable trees. Leaves, flowers and pods from these trees were available for most part of the year. Some of the tribals who sold the drumsticks in market could get 150 to 200 rupees from one tree. This proved that drumstick was also an income-generating tree.

All the above plants could be propogated by means of seeds, cuttings or grafts. Such low cost technologies lie within the reach of poor tribals and hence the programme of Nutrition Gardens was self replicable.

The diet of tribal families which earlier comprised only of cereals, now also contained vegetables. They also started consuming fruits. Nutrition Garden was a self-help programme for combating malnutrition among the weaker sections of the society particularly the tribals.

Dr. Charles Janeway World renowed paediatrician and Dean of the Harvard Medical School visited India to attend an International Paediatric Congress. His daughter Miss Anne Janeway worked with me in training of the Amercian Peace Corps at Brattleboro in Vermont State as well as at the Kosbad Institute. She requested her father to visit the Kosbad Institute to see the work for tribal upliftment. We showed the project of Nutrition Gardens to Dr. Janeway. He told me that there were millions of children in India who suffered from chronic malnutrition, and therefore, production and consumption of nutritive foods was of utmost importance. He was particularly happy to see the two vegetable trees of Moringa and Agati which supplied both leaves and flowers rich in vitamin 'A'. He advised me to concentrate on the propogation of these two trees as they were drought resistent and could be grown in rainfed and dry areas. He also advised me to introduce orange colour vegetables and fruits as carrot, pumpkin, papaya and mangoes rich in 'A' vitamins. We followed his advice and prepared a nursery of orange coloured vegetables and fruits plants. The seedlings were distributed to the tribal families who planted them in their backyards.

The officers of UNICEF from all over the World visited the project of Nutrition Gardens. The New York office of UNICEF prepared a set of slides on the project for distribution.

Education in Child Care

After receiving training in pre-school education my wife Meena started a creche and *balwadi* on the farm of the Kosbad Institute in 1976. Tribal children from Varaipada and Lilakpada were coming to this *balwadi*. The mothers were bringing their infants to the creche. This had greatly facilitated working women. We prepared a hut for conducting the *balwadi* and creche. We named it 'Parnakuti'. It was prepared from bambooes and the roof was covered with rice straw and leaves of teak. It resembled the hut of tribals and so the children felt it at home. It was open from all sides and so there was plenty of fresh air and light inside the hut. A circular platform was prepared round the hut which was plastered by red earth and cow dung. This was used as the playground for the children. We also constructed a children's toilet and connected it with biogas plant. A small water tank was constructed for bathing children.

The tribal children were not bathing everyday and many children had skin diseases. We therefore, developed the habit of daily bath among the children. The tribal children were using green leaves as toilet papers. They did not clean their hands with water. This often caused gastro-intenstinal infections. We showed the children how to wash their hands. The nails of the children were also regularly cut with nailcutters. Due to formation of habits of cleanliness, the incidence of ailments among the children reduced significantly. Immunization was carried out regularly for the children and infants from the *Balwadi*.

We prepared a nutrition garden near the *balwadi* and planted green leafy vegetables, beans, okra, egg plant and cucurbits. Similarly quick growing fruit trees like papya and bananas were also planted. We were preparing *pulav* (rice mixed with vegetable) for mid-day meals of the children. The vegetables from nutrition garden were used for this purpose. We found that if a nutrition garden is attached to the *balwadi* it helps to implement the programme of mid day meals to the children. The children used to play near the nutrition gardens. They were picking fruits and vegetables. We prepared a small tank in which fish was reared. The children were playing the game of fishing. We also kept chicks, rabbits and lambs. The children were feeding them. This aspect of 'learning while playing' was found to be very useful in formation of creativity among the children.

The mothers who kept their infants used to come in the noon to collect them. We gave them simple training in planting vegetables, fruit trees, budding and grafting. Simple receipes of vegetables were also shown to them. We used to demonstrate the preparation of weaning food for children. The training was followed by supply of nucleus material like plants and grafts. We found the atmosphere of *balwadi* was most congenial for giving education to mothers in child care.

Introduction of Soyabean

Tribals in Thane district grew pulses on uplands. However the area under such uplands was very small and so the total produce was sufficient to feed the family for two to three months only. Lack of pulses in the diet often led to deficiency of proteins amongst the tribals. This seriously curtailed the ability of tribals to put in hard labour required for agricultural operations.

I attended the conference of Krishi Vigyan Kendras held at Indore in the year 1977. We were taken to see the cropping system practiced in Malwa tract of Indore region. The soils here were deep and black enriched by assured rains. Cotton was the traditional crop of this region. However, soyabean has now attained dominance in this region. Soyabean is used chiefly for extraction of oil. Large numbers of processing plants have been established in the region due to the large acreage under soyabean crop. Soyabean meal has demand both in the domestic and export market. Due to high protein content soyabean products are becoming popular for human consumption too.

While returning from Indore I carried some seeds of soyabean. We had sown the seeds on the uplands of tribal farmers in the *Kharif* (rainy) season. We also prepared a demonstration plot on the farm of the Kosbad Institute. Since this was a new crop for our region, we followed the practices as carried out in Madhya Pradesh. The crop was ready within hundred days. We arranged a demonstration of soyabean at Kosbad Institute, attended by a large number of tribal farmers. These farmers were also taken to the fields of tribals who grew soyabean. We showed them different recepies of soyabean like preparation of bread by mixing of soyabean flour. After the demonstration a sample of soyabean seeds was supplied to each farmer.

During the next Kharif season the tribal farmers grew soyabeans on their uplands and on bunds of rice fields. Soyabean crop was ready for harvest before the rice crop and so was found to be an appropriate mixed crop.

Demonstrations of soyabean recipes were also held for tribal women in the Institute's home science laboratory, and women were trained in growing soyabean. The soyabean crop thus became popular among the tribals. Since soyabean has high protein content its introduction among the weaker sections helped in overcoming protein deficiencies.

5

FARM INNOVATIONS

Supplementary Food through Tuber Crops

The tribals are very fond of tubers. The tribal women go to the forest for collecting tubers. They carry a bamboo basket, a crow-bar and a pickaxe. They have inherited the knowledge about the location of tubers in the forest areas. They consume the tubers either by boiling or by roasting. They subsist on the the tubers whenever they face food shortages. Many a times, these tubers are very bitter and so they are cut into pieces and kept in ash. Next day they are boiled in water and eaten.

There is one tuber called *Bhadurli* which is a favourite of the tribals. It grows in the grasslands and is ready for harvest after the rains are over in October. The tribal children who go to the meadows to graze their cattle dig these tubers. They roast these tubers by burning in coarse grass. Tribal children are very fond of these tubers.

The tribals believe that the tubers are very medicinal and one must consume them. This tradition of consumption of tubers has been there for generations.

We took up a programme to introduce good quality tubers among the tribals. We selected sweet potatoes, and the variety "Ceylon Clusters" was considered suitable for the purpose. This variety is high-yielding and the quality of tubers is also good. We prepared a nursery of sweet potato vines on the farm of the Agricultural Institute. The nursery was prepared in the month of May so that the distribution of vines could be possible in second week of June when the rains start in the region. The tribals were given training for preparation of land for planting of sweet potato vines. They prepared ridges and furrows on their uplands. They also applied farm-yard manure and compost. Application of organic manure helped to improve the water holding capacity of the soil. Similarly it also improved the texture of the soil thereby facilitating good growth of the roots. The organic manure also supplied necessary plant nutrients. All these things were explained to the tribals during their training. The tribals were invited to the farm to take the sweet potato vines. We gave them demonstration on the planting of vines. After the training every tribal carried the vines according to requirement. For the first time in their lives they planted sweet potatoes in their lands. The tribals took after-care of the crop like weeding and earthing up. They consumed the sweet potatoes either by boiling or by roasting. Since sweet potatoes are non-perishable, they lasted for about 1-1/2 months. The tribals were consuming them as and when required.

We prepared a community nursery of sweet potatoes for supply of vines to the tribals. This nursery was raised in summer, so that the vines could be supplied at the onset of rains in June. We distributed about fifty cartloads of vines to the tribals of Thane district. The tribals planted them on their uplands. At that time there was "Grow More Food Campaign" in the country. Pandit Jawaharlal Nehru Prime Minister of India made an appeal to the farmers to produce more food. Acharya Bhise desired that we should write a letter to the Prime Minister that we have responded to his appeal and distributed fifty cartloads of sweet potato vines to the tribal farmers. We would like to enroll ourselves as soldiers on food front. Pandit Jawaharlal Nehru liked this letter very much and he gave this letter for publicity to all the newspapers in the country. The *Times of India* in its Bombay edition gave a news on front page saying " Nehru Lauds Bordi Example". He commended the efforts done by the Kosbad Institute and appealed to all educational institutions in the country to follow it.

This shows that sweet potato could be grown successfully as a rain-fed crop in the tribal areas. However, community nurseries will be required for supply of vines during the beginning of June when the rains start. Such community nurseries could be started where there is a facility for irrigation. Growing of sweet potatoes was found to be an effective programme for elimination of starvation among the tribals.

Japanese Method of Rice Cultivation

Shri Harishchandra Patil, Director of the Kosbad Institute, and I attended a demonstration of rice crop at Kora Gramodyog Kendra, Borivali, Bombay in October 1949. Coir strings were tied all around the rice crop and in between the lines also at a height of about 3 feet. The rice crop where strings were tied was standing erect. While in the adjoining field rice crop completely lodged on the ground as no strings were tied.

Shri Pran Lal Kapadia, Director of the Kora Gramodyog Kendra had organised this demonstration. He said that this method was being followed in Japan. Shri Kapadia had stayed for several years in Japan and observed this method of tying strings. This demonstration was attended by the Chief Minister of Bombay State, Shri Bala Saheb Kher, Finance Minister, Shri Vaikunth Lal Mehta and Minister for Local Self Government, Shri Anna Saheb Vartak. Shri Kapadia told in his speech that we should adopt the Japanese method of rice cultivation in India. It would boost our rice production and help attain self-sufficiency in food production. He felt that the Government should depute Agricultural Experts like Shri Harishchandra Patil to study the Japanese method of rice cultivation.

I was regularly visiting Kora Gramodyog Kendra with Shri Harish Chandra Patil and so I got acquainted with Shri Pran Lal Kapadia. He gave me a book titled *Birds Eye View on Agriculture in Japan*, which had a very good description about rice cultivation in Japan. Sowing seeds on raised beds, interculturing between two rows, tying the strings were some of the salient features of the Japanese method of rice cultivation. There was one important sentence in the book which said "Agriculture in Japan is looked upon as Horticulture, every single plant of rice is treated as a fruit tree". Japanese farmers use the green manuring plant called "Chinese Milk Vedge". The fertilisers are given not in a single dose, but in split doses. After transplanting of rice crops, interculturing is done with a hand operated tiller. Strings are tied in the field to prevent the rice crop from lodging. The rice crop is threshed by a pedal operated tool. These were some of the features described in the book.

In 1951 the Government of Bombay sent a delegation to Japan to study agriculture and village industries. Shri Harishchandra Patil, Director of the Kosbad Institute, was included in this delegation. Shri Patil studied the Japanese method of rice cultivation by actually working for 6 months on a farm in Hamamatsu County. It helped him acquire skills of the Japanese method of rice cultivation.

While Shri Harishchandra Patil was in Japan, I experimented with the Japanese method of rice cultivation on the farm of the Kosbad Institute in 1951. I followed the instructions from the book which Shri Pran Lal Kapadia had given me. Shri Harishchandra Patil was also sending me instructions from Japan. A rice variety known as "Zenia-31" was selected for this purpose. Organic manures and fertilizers were added in adequate doses before transplanting. Transplanting was done in straight rows. This was the first time that transplanting was done in such a way. Interculturing was done by the help of a pick axe. Weeds were removed regularly. Cattle dung mixed with urine was also applied after every fifteen days. Shri Vishnu Thanekar, a student from the Kosbad Institute was doing all these operations regularly. Due to the application of this organic liquid manure the growth of rice crop was very vigorous. Strings were tied along the borders as well as in between the lines which prevented lodging of the crop. The whole field was full of panicles and looked very impressive.

In the month of October we arranged one demonstration of this experiment. Shri Harishchandra Patil had also returned from Japan. We invited Justice Sham Rao Tendulkar who was a Judge of the Bombay High Court. Shri Tendulkar was very fond of agriculture and he used to visit Bordi in connection with Topiwalla Agricultural Trust. Shri Tendulkar was greatly impressed with the rice crop cultivated according to the Japanese method. He appealed to the farmers to adopt this method. Shri Harishchandra Patil narrated his experiences in Japan. Selection of good quality seeds, dipping the seeds in salt water, treatment of seeds with fungicides, sowing the seeds on raised beds, application of green and organic manures before transplanting, adequate use of fertilizers, transplanting in straight rows, interculturing, plant protection and tying of strings to prevent crops from lodging were the features of Japanese method of rice cultivation. The farmers who were assembled for the demonstration were greatly impressed with Shri Patil's observations.

During the Kharif season of 1952, demonstrations of the Japanese method of rice cultivation were arranged at the Kora Gramodyog Kendra, Borivali and the Kosbad Institute. Pandit Jawahar Lal Nehru, Prime Minister of India attended the demonstration at Kora Gramodyog Kendra. He was greatly impressed by this method and said that his method should be advocated all over India. Dr. Panjab Rao Deshmukh, Minister of State for Agriculture, Government of India attended the demonstration at Kosbad Hill. He described Kosbad Hill as the 'Mecca of the Japanese method of rice cultivation'.

The Ministry of Agriculture had organised a Conference of Directors of Agriculture at New Delhi. Pandit Jawahar Lal Nehru presided over this Conference. Shri Pran Lal Kapadia and Shri Harishchandra Patil were specially invited to narrate their experiences about the Japanese method of rice cultivation. One of the Directors of Agriculture told the Conference that there is nothing new in the Japanese method and we know all these aspects. On hearing this Pandit Jawahar Lal Nehru was very much annoyed and asked that Director, "if you were knowing all these things, why did you not follow? The country is deficient in food production and we have to beg for food from other countries." He directed the Minister of Agriculture to see that this method was replicated all over the country. Dr. Panjab Rao Deshmukh, Minister of State for Agriculture took great initiative to propogate this method all over India. The Ministry of Agriculture also published one book namely *India's Rice Revolution*. Shri Pran Lal Kapadia and Shri Harishchandra Patil visited all States under the auspicious of Gandhi Memorial Trust and gave demonstrations of the Japanese method of rice cultivation. These demonstrations were very effective and were attended by the Chief Ministers of respective States. All the States took great interest in the adoption of this method.

We arranged one demonstration at the Kosbad Institute in 1952. This was attended by Shri Bhau Saheb Hiray, Minister for Agriculture, Government of Bombay. He was very much impressed to see the rice crop grown by this method. He invited Shri Harishchandra Patil to accept the responsibility of propogating this method all over Bombay State. Shri Patil was appointed Officer on Special Duty for the Japanese Method of Rice Cultivation. Shri Patil demonstrated this method by actually working in the field. This impressed the farmers and they felt that he was a real farmer demonstrator. It created a great impact on the farming community and they adopted this method of rice cultivation. In this way Shri Pran Lal Kapadia and Shri Harishchandra Patil were greatly responsible for the introduction of the Japanese method of rice cultivation in India. *Readers Digest,* the world renowned magazine had published one Article on the Japanese Method of Rice Cultivation in India. It stated that due to this method a revolution is taking place in India's rice cultivation. It gave great compliments to Shri Pran Lal Kapadia and Shri Harishchandra Patil for their contributions in introduction of this method. The Japanese method of rice cultivation has now come to stay all over India.

Introduction of High Yielding Varieties of Rice

It was essential to step up supplies of food grains keeping in view the growing population of our country. In 1965, Government of India took up a programme to introduce high yielding Taichung native-1 variety of rice. This variety was evolved by agricultural scientists in Taiwan. The variety was semi-dwarf and responded well to fertilizer application. Its yield was nearly double than the local varieties. The Ministry of Agriculture, Government of India arranged to send us a seed sample of this variety. We conducted experimental plantation on the farm of the Kosbad Institute. The crop flowered in the first week of September. The whole crop was erect and did not lodge down on the ground. We arranged a demonstration, attended by large number of farmers from all over Thane district. The farmers had never seen such semi-dwarf and erect variety of rice and demanded seeds for cultivation in the subsequent kharif season.

Accordingly, we arranged for seeds from the Ministry of Agriculture and distributed them among the farmers. The farmers who grew it, found that due to its semi-dwarf character it did not lodge on the ground like the local varieties. This variety of rice was adopted by the local farmers permanantly.

In the Kharif season of 1967, the Ministry of Agriculture sent us a sample of new rice variety I.R.-8. This variety was developed by scientists at the International Rice Research Institute in Philippines. It was a cross between two varieties viz. Peta and De-woo-Jung. Due to its high yielding nature it was known as miracle rice. We arranged a demonstration of this variety and invited Shri Vasantrao Naik, Chief Minister of Maharashtra to be the Chief Guest. Farmers from all over Konkan attended this demonstration and were greatly impressed by the performance of this variety. Shri Naik instructed the officers of the Agriculture Department to bring large quantities of seeds for the next Kharif season. Accordingly, these seeds were distributed by the Zilla Parishad in Thane District. Meetings of farmers were arranged by us in March and April in every Tehsil to explain the importance of this variety of rice. We got very good response from farmers. During Kharif season of 1968 the area under I.R.-8 increased considerably. We found that for the introduction of high yielding varieties holding demonstrations and farmer's rallies are of primary importance.

After some years the scientists at the All India Coordinated Rice Improvement Project at Hyderabad evolved a new variety of rice called 'Jaya'. Dr. Shastri who was the coordinator of this project sent a sample of seeds of Jaya for making experimental plantation at the Institute. The stand of this crop was even better than I.R.-8. We requested Dr. Shastri to visit Kosbad Institute to see the crop. Dr. Shastri came alongwith Dr. Freeman, a scientist at the International Rice Research Institute. Both found that the crop was free from bacterial blight, a disease prevalent in Andhra and other States. They were of the opinion that seeds from such disease free crop should be multiplied for subsequent uses. We decided to hold a demonstration and invited Shri Vasantrao Naik, Chief Minister of Maharashtra to be the Chief Guest. Farmers from all over Maharashtra attended this demonstration. The farmers were impressed to see that the crop was resistant to diseases. As a result, in the Kharif season of 1968 there was large coverage under 'Jaya' variety.

Most of the tribals whose lands were on hillslopes could not grow Jaya as it was a mid-late variety. They required an early variety of the crop. Shri M.S. Pawar, who was working as rice expert of Food and Agricultural Organization (FAO) of United Nations in Guyana wrote to me in a letter that the farmers in that country were growing an early rice variety called

'Blue-Belle'. This variety matured within hundred days and was high yielding. Due to its early nature the farmers could grow three crops in a year. He sent me a sample of seeds which was planted in an experimental plot. The crop flowered in 70 days and was ready for harvest in hundred days. Its earheads were long and each had 250 to 300 grains. This variety was evolved by Dr. H.M. Bechel at Beamount Experiment Station in Texas in U.S.A. Dr Bechel also bred I.R.-8 at the International Rice Research Institute in Philippines. I requested scientists at Beamount Experiment Station to send me pure seeds of 'Blue-Belle' and they were kind enough to supply a kilogram of the pure seed. These seeds were subsequently multiplied by us. The tribals in the area demanded the seeds of 'Blue-Belle' as the crop was suitable for their uplands. We prepared mini-kits of seeds and distributed them amongst the tribals. The early maturing nature of this variety satisfied the tribals. We were able to introduce Taichung native-1, I.R.-8, Jaya and Blue-Belle varieties among the farmers of Maharashtra in this manner.

With a view to introduce high yielding varieties of rice among the tribals, we undertook programme of community nurseries of rice. We prepared these nurseries in summer with irrigation water, so that the seedlings were ready in the third week of June. These months receive good rains and it is the most appropriate time for transplanting. The tribal women came from long distances to take the seedlings. Due to early transplanting, the growth of their crops was very healthy and they received better yields. We found that community nurseries of rice play an important role in the introduction of high yielding varieties among the weaker sections.

We invited Dr. M.S. Swaminathan, Director General, Indian Council of Agricultural Research (ICAR) to see the high yielding varieties programme among the tribals of Kosbad Hill. We took him to a tribal farmer Radhya Dhak who grew Jaya variety. He told Dr. Swaminathan that he had now become self reliant in food production. Previously he had to subsist on gruel prepared from rice crushings. We found that high yielding varieties of rice was the most effective programme for overcoming hunger among the tribals.

Three Crop Cycle

The tribal farmers in the areas surrounding Kosbad Hill who had facilities of wells and pumps were growing high yielding varieties of rice like Jaya and Ratna. Due to assured irrigation it was possible for them to grow these varieties. After the harvest of rice crop in October they had introduced high yielding varieties of wheats like *Kalyansona* and *Sonalika*. The wheat crop was harvested in last week of March. However, the land remained fallow in April and May.

We were thinking of introducing short duration crop in summer. I contacted scientists of the Indian Agricultural Research Institute, New Delhi. When I went there in March 1971, they told me that they had developed a short duration *Mung* (Green Gram) variety which would be ready within sixty days. Further, this variety was suitable for the summer season. They named this variety 'Baisakhi Mung', and gave me a seed sample of about 2 kilograms. which I carried to Kosbad Hill.

The wheat crop was harvested in the last week of March. The seeds of *Mung* had to be sown immediately. We did not have time for ploughing and other land preparations. We decided to use this field as seed bed for *Mung*. The *Mung* seeds were dibbled in between lines of wheat stubbles. Basic dose of phosphatic fertilizer was applied before sowing. Watering was done immediately after sowing of seeds. Four light waterings were given after intervals of fifteen days. The crop was ready for harvest in the last week of May. It was dried for four days in sun and then threshe

The weaker sections's like tribals who have protein deficiency got nutritious diet through consumption of *Mung* beans. The cattle received proteinous fodder from the vines and leaves. *Mung* being a leguminous crop fixed nitrogen thereby enriching the soil. It was therefore the most appropriate crop for rotation after rice and wheat.

Since Baisakhi Mung was a new crop of the region we arranged demonstrations for farmers from Thane district. All the farmers were impressed to see this short duration summer crop and they expressed intentions to introduce this crop during the following summer. We called for large quantities of *Mung* seeds from National Seeds Corporation. The Thane Zilla Parishad organised distribution of seeds through Panchayat Samitis. Large numbers of farmers from Thane district grew the *Mung* crop for the first time during that summer.

In this way we could succeed in our experiment of evolving a three crop cycle of rice, wheat and *mung*. We were able to introduce this three crop cycle among the small farmers and tribals, making them self reliant in cereals and pulses. Besides, the small farmers and tribal got year round gainful employment on their lands.

Introduction of Glyricidia for Green Manure

In the Konkan region, rain commences from second week of June when seeds of rice are sown. The transplanting of rice is done in the month of

July. The fields are vacant for about one month during this period. We, therefore, made some experiments to introduce green manure crops. We had sown sunhemp and Dhencha during this period. The sunhemp could not grow due to water logging in the fields. Dhencha had some tolerance towards water logging. However, both these crops did not give enough biomass required for green manure.

We came to know that in the southern states of India, leaves of glyricidia are used as green manure for rice crop. This is a very fast growing tree and gives about 100 kilograms of green leaves per year. We called for seeds of this plant and prepared its seedlings in polythene bags. These seedlings were planted on bunds of fields and also on uplands. The growth of the seedlings was very fast. During one year they attained height of 6 feet. The tree did not require any watering in summer which showed their drought tolerance. In the month of March the trees bloomed with pink coloured flowers. Pods were formed from these flowers. We collected the seeds from these pods in May. We prepared seedlings in polythene bags and distributed them among the surrounding tribal farmers. They planted them on the bunds of their rice fields. Some of them also planted on their uplands.

We arranged demonstrations of green manure on field of a tribal farmer. A tribal farmer who grew glyricidia on his land first chopped the branches and put the leaves in soil at the time of puddling. All the farmers who assembled for the demonstration practiced the method. Transplanting of rice was done in the field where green manuring was done. The rice crop in green manure plot was healthy and gave higher yields. The tribals were convinced about the importance of green manure and they adopted this practice. It was a low cost technology and hence was within the reach of poor tribals.

One day, Shri B.K. Chaugule, Secretary to Government of Maharashtra, Agriculture Department visited Kosbad Institute. We showed him a nursery of glyricidia seedlings. We also took him to the surrounding villages where tribal farmers were practicing green manuring with glyricidia. He was convinced about this green manure tree and desired that this tree needed introduction not only in Konkan but all over Maharashtra. Shri Chaugule had a background of agriculture as his father received 'Krishi Pandit' award for achieving the highest yield of *Jowar* (sorghum) in an All India competition. He told us that the glyricidia seedlings prepared at Kosbad Institute be distributed to farmers from ten districts. On his return to Bombay he issued a circular to ten Zilla Parishads (District Councils) to secure seedlings from Kosbad Institute. Accordingly, the Zilla Parishads carried about fifty thousand seedlings and distributed them among small and marginal farmers. The farmers who introduced glyricidia have started using glyricidia leaves for green manuring.

The soils in tropical countries like India require constant replenishment with organic manures. However, the traditional organic manures like farmyard manure (cattle manure) and compost have become scarce. Green manure is one of the alternative to overcome the shortages of organic manures. Glyricidia a fast growing tree was found to be most suitable tree for green manure.

6

CROP INTRODUCTIONS

Experiment of Growing Groundnut in Summer

The farmers in Thane district who had irrigation facilities were growing vegetables after the harvest of the rice crop. Most of the farmers sold these vegetables in the nearby townships. Some farmers were sending their vegetables to markets in Bombay. However, since all these vegetables were ready at the same time, their prices would crash in the market. Naturally the farmers had to incur great losses. The farmers whose lands were in the interiors and did not have facilities of rail or road transport were facing more difficulties. They had to carry their vegetables in bullock carts to long distances. Since all these vegetables were perishable they had to bear great losses if they were not sold in time.

There was a need for an alternate crop to vegetables, especially where there were no facilities of transport. I decided to make experiments in this direction.

This is a story of 1957. A friend of mine, Dr. Madhukar Rane visited our Institute at Kosbad Hill. He was a lecturer in Plant Pathology in the College of Agriculture at Pune. I discussed the possibilities of growing various crops under our agro-climatic conditions. We thought that there was a possibility of growing groundnut in the winter season as the winter in Konkan is not severe. We therefore decided to make an experiment of cultivation of groundnut. Dr. Rane sent me seeds of an improved groundnut variety viz. S.B.-11. After the harvest of the rice crop we ploughed the land in October. Farmyard manure and compost was added to the soil. The organic manure helped to make the soil porous. We then prepared flat beds and seeds were sown in them. We gave light watering to the beds. We followed necessary practices like interculturing, application of fertilizers and plant protection. After four months the crop was ready for harvest. The growth of groundnut pods was very good. The oilseed specialists at Jalgaon expressed satisfaction after seeing the crop. In 1958 we planted this crop in summer i.e. between February to May. This crop performed better than winter crop. This experiment showed that groundnut could be grown both in winter and summer in Konkan.

The Department of Agriculture of the Government of Maharashtra started a scheme for cultivation of groundnut in Konkan during the summer season. The scheme was strengthened in the year 1980 and all the four districts of Konkan were covered under this Scheme. The farmers who had access to small irrigation wells and lift irrigation schemes, grew the groundnut successfully. Since groundnut is a non-perishable crop, the farmers in the interior areas who did not have any transport facilities found it suitable. In some of the villages some entrepreneurs had set up oil expellers. The farmers who grew groundnut have become self reliant in edible oils. Similarly, groundnut, being a leguminous crop, enriched the soil through nitrogen fixation. The plants of groundnut were used as fodder. The country was in need of edible oil self-sufficiency. In that direction introduction of a summer groundnut crop helped the farmers to attain self-sufficiency in oil seed production.

Experiment with Laxmi Cotton

I had to go to Dharwad in Karnataka in 1958 to write my thesis for the Master of Science degree as my Professor, Shri U.K. Kanitkar was working as Plant Physiologist of cotton at Dharwad. Prof. Kanitkar used to take me to show cotton research stations in Dharwad district. Once he took me to Gadag which is a big market place of cotton. 'Laxmi' variety of cotton was grown on a large scale in that area and this variety was brought for sale in the market. This variety was in great demand.

Once I went to see the cotton cultivation in Malnad area of Dharwad district. The soils were red laterite and the climate was humid. The agro-climatic conditions were similar to those of Konkan. I enquired with Prof. Kanitkar about the important features of 'Laxmi' cotton. He informed me that this variety belongs to 'Indo-American' or 'hirsutam' type which needs humidity when the opening of bolls takes place. He also informed me about the cultivation practices of this variety.

While returning from Dharwad I brought some seeds of 'Laxmi' cotton and decided to make an experiment at Kosbad Hill. It was the month of October. The early variety of rice was just harvested and we were planning to raise a bean crop on residual moisture. We selected one such field and ploughed it while it was in optimum moisture condition which is locally called as 'wapsa'. The seeds of Laxmi cotton were sown at one meter distance in straight rows. Since there was good moisture in the soil all the seeds germinated well. We sprayed insecticides to protect the seedlings from leaf eating caterpillars. My friend Dr. Madhukar Rane who was working in the plant-pathology division of the College of Agriculture at Pune had given me valuable guidance about plant protection.

The growth of the cotton plants was quite fast. We observed that the plants did not grow tall but remained dwarf as there were more branches. After two months, flowering started. We took appropriate plant protection measures against boll-worm attack. The opening of the bolls was good. The leaves remained green till harvest of cotton. We obtained 130 kgs. of cotton per acre from that experimental plot.

I requested Prof. Kanitkar to visit Kosbad Hill and see this experiment. He was surprised to see this crop grown simply on residual moisture. When he saw that the leaves of cotton plants were green even in summer he concluded that this was due to high moisture retention in the soil. He told us that in Karnataka the leaves become dry at the time of harvesting of cotton. Due to this the trash of dry leaves gets mixed with cotton affecting its quality. Here due to green leaves we obtained clean cotton. He observed that the good opening of bolls was due to good humidity in the air. The Konkan tract remains humid throughout the year, due to its proximity to the sea. He further told that if the climate is dry at the time of maturity of bolls then the bolls do not open. In such conditions the closed bolls are harvested and are opened by hand labour.

We sent the sample of Laxmi cotton for testing to Cotton Technological Laboratory at Matunga in Bombay. We were informed that the quality of this cotton is as good as cotton in Karnataka. The results of this test boosted our confidence in growing 'Laxmi' cotton. Next year in the rabi season of 1959 we planted two acres of Laxmi cotton after the harvest of the rice crop. The growth of crop was good and we obtained an yield of seed cotton upto 180 kgs. per acre. In the 1960 rabi season we planted six acres of Laxmi cotton and obtained yields upto 105 kilograms per acre while in 1961 season it was planted on four acres and the yield was 161 kilograms per acre.

In 1961 there was a meeting of All India Cotton Development Conference at Bombay where Prof. U.K. Kanitkar and myself submitted a paper entitled "Possibilities of Growing American Cotton Laxmi in Rice Fallows Without Irrigation in Coastal Districts of Maharashtra State". In this paper we stated that the quality of cotton was similar to that in Karnataka. Our paper was appreciated by all the participants. This paper was published in a scientific journal, *The Indian Cotton Growing Review* in January 1964.

We arranged a demonstration of Laxmi Cotton at Kosbad Hill and invited Shri Vaikunthlal Mehta, Chairman, Khadi and Village Industries Commission. Shri Pran Lal Kapadia, Secretary of Khadi and Village Industries Commission also attended the demonstration. Both of them were greatly impressed to see that cotton could be grown in Konkan without irrigation as a second crop after rice. They told us that Khadi and Village Industries Commission would be ready to support a research project for growing cotton in Konkan. Accordingly, we undertook this project in Thane, Raigad and Ratnagiri districts. We arranged demonstrations to show that this cotton could be grown as second crop after harvest of rice on residual moisture. As there was no market in Konkan we were sending this cotton to Navasari in the Surat district of Gujarat. Some of the farmers grew this cotton very successfully but as there was no nearby market they could not sell their produce. Even though this cotton grew well its cultivation could not be extended due to lack of marketing facilities. However, we demonstrated that a valuable crop like cotton could be grown on residual moisture after harvest of rice crop.

Asparagus

While studying at the Kansas State University in the U.S.A., I got an opportunity to observe the cultivation of vegetables of the temperate region. Asparagus *(Asparagus officinalis)* is one such crop. It is used for preparing soup and such other recipes. It is considered as a delicacy and hence is in great demand in the market.

Asparagus plants bear seeds which are of black colour. The seeds are sown on raised beds. When the seedlings are about six months old they are uprooted from the beds and transplanted in the field. Since this crop stays for about eight to ten years in a field, good preparatory tillage is required. Organic manure is also added in large quantities.

When the winter starts there is a snow fall. The plants and leaves begin to dry and fall to the ground. During winter the plants become dormant. When the spring begins, the asparagus crowns start sprouting. These sprouts are called "Spears". The spears are harvested early in the morning. A knife with a long handle is specially prepared for harvesting asparagus. The harvested spears are kept in a cold chamber for pre-cooling, which helps to increase the shelf life of spears. The spears are tied in small bundles, packed in polyethelene bags and sent to the market.

When I returned to India in July 1964, I brought some seeds of asparagus. The variety was Mary Washington. The seeds were certified by the Plant Quarantine Department of the United States Department of

Agriculture. I decided to carry out the experiment of introduction of Asparagus. The seeds were sown on raised beds and when the seedlings were six months old they were transplanted to a field. Good preparatory tillage was done in this plot as the crop is to stay for a number of years. It consisted of deep ploughing, application of organic manure and laying of good drainage. The root crowns of the asparagus seedlings were planted deep in furrows. Such deep planting helps to hold the roots inside the soil. It was observed that exposure of roots outside soil adversely affects the growth of asparagus plants. Protection of roots was therefore very necessary. Irrigation was given soon after planting. The plants grew vigorously in the rainy season. We expected that during winter the asparagus plants would become dormant. However, due to warm humid climate there was no dormancy and the plants continued to sprout and give spears. We marketed these spears locally. We gave demonstrations about preparation of soup and other recipes of asparagus. Asparagus soon became popular among local people.

The American Peace Corps Volunteers used to come to the Kosbad Institute, for training. The American Consul-General in Bombay, Dr. Daniel Braddock visited the Institute to meet these volunteers. We prepared soup of asparagus for lunch. He was very surprised to see the soup prepared from fresh asparagus. Mrs. Braddock, wife of the Consul General told me that they had an American Women's Association in Bombay and they would be happy to buy asparagus from our Institute. They did not get fresh asparagus in Bombay. We agreed to their proposal. Accordingly we were harvesting asparagus spears early in the morning and sending them to Bombay by morning train. Mrs. Braddock made arrangement to distribute them to the members of the American Women's Association. The American women greatly appreciated the fresh asparagus. Uptil now they had to depend on tinned asparagus from Europe or the United States. Once I asked Mrs. Braddock about the quality of our asparagus and she said that there was more fibre compared to asparagus in the United States. However, she was happy as she could get fresh asparagus.

I used to write regularly to Dr.Hanna, Professor of Vegetable Crops at University of California to seek his guidance about growing asparagus under tropical conditions. I wrote to him that due to warm weather there was no dormancy of the crop and that it continued to give spears throughout the year. I have also informed him about the higher fibre content in the spears. Dr. Hanna had studied this crop from many countries and had a wide experience. He informed me that we should harvest the spears regularly early in the morning and refrigerate them immediately.

This would increase shelf life. He also wrote to me that we should harvest the spears throughout the year as there was no dormancy. We should apply more organic manure to maintain the physical condition of the soil. He also suggested laying of drainage to avoid water logging during heavy rains. He further wrote that in the United States, the crop is rotated after eight to ten years as the crop has rest period of about five to six months during winter season. But in India since the crop does not have any rest period the soil is exhausted early. He therefore suggested a rotation of four to five years. I was in regular touch with Dr. Hanna who gave me valuable guidance about asparagus.

Shri Murlidhar Dhoot a progressive farmer from Aurangabad was impressed to see Asparagus and decided to introduce this crop on his farm at Gangapur. We gave him some seeds to start cultivation. Later he called for more seeds of Mary Washington variety from U.S.A. He grew this crop very successfully and sent fresh spears regularly to Bombay for sale. He gave some spears for canning also. Two farmers viz. Shri Bhalchandra Patil and Shri Makarand Churi from Bordi are cultivating asparagus successfully on their farm at Jalwai near Bordi. They developed various agronomic practices suitable under the agro-climatic conditions of Konkan. Their fresh asparagus spears are in geat demand in the Bombay market.

Asparagus is a crop of the temperate region. We made efforts to introduce this crop under tropical conditions. It took years to develop location specific technologies. I received great satisfaction to see that our efforts bore fruitful results.

Bringing Wheat Cultivation to Konkan

Shri Lal Bahadur Shastri, former Prime Minister of India gave the call of 'Jai Jawan Jai Kisan' in 1965. He paid great compliments to soldiers and farmers for their contribution.

The country was deficit in food production and needed move towards self sufficiency. We decided to respond to the call of the Prime Minister by growing more food crops.

In the Konkan region rice is grown between June and October. That is the only food crop grown in the region. The harvest of rice was followed by growing vegetables. We decided to grow wheat during winter. It was believed that due to higher temperatures in Konkan, wheat could not be grown.

We obtained wheat seeds of improved varieties from the wheat research station at Niphad in Nasik district. We made land preparations and sowed the wheat seeds carefully. We followed all the practices recommended by the wheat specialist. The growth of the crop was very good. We invited the wheat specialist to see the crop. He was surprised to see the excellent stand of the crop and remarked that it was comparable to the wheat crop grown in traditional areas of Maharashtra.

The experimental plot was given the name of Lt. Dilip Gupte who lost his life in the Indo-Pak war. We arranged a demonstration of this wheat plot. We invited Dr. Hemchandra Gupte, father of Lt. Dilip Gupte as Chief Guest. Dilip's mother was also present on the occasion. They were happy to see this experiment dedicated in name of their son who had given his life for the country. The farmers who assembled for the demonstration were impressed to see that wheat could be grown in Konkan. They decided to grow wheat on their lands during the next winter season.

We arranged a meeting of the office bearers and officers of the Zilla Parishad and Panchayat Samitis at the Kosbad Institute, in September 1966. A programme was prepared to introduce wheat cultivation in Thane district in the coming winter season. A target of ten thousand acres was set. Arrangements for procurement of good quality seeds from Nasik district were made. It was decided to organize training camps of farmers in each taluka in September and October. We took the responsibility of organising such training camps. We gave demonstrations of land preparation, sowing seeds and after care of wheat crop. The farmers gave good response to our appeal and attended the camps. In November the farmers sowed wheat on their lands. The area under wheat was about twelve thousand acres which exceeded the target. It was the first time that they grew this crop. There was great risk in growing a new crop but they took this risk in the interest of the nation. We were continuously touring the district and giving guidance to farmers. Most of the farmers got yields ranging from 800 to 1200 kilograms per acre. In this way wheat was grown for first time on such a large scale in Konkan.

The country imported seeds of Mexican varieties for distribution in Punjab, Haryana and the Western part of Uttar Pradesh. I wrote a letter to Dr. M.S. Swaminathan, Director, Indian Agricultural Research Institute, Delhi requesting him to send me a sample of seeds of Mexican varieties for experimentation. Accordingly he sent a sample of Sonora-64. We cultivated this variety on the farm of the Kosbad Institute. We found that it was an early and dwarf variety. The panicles were bigger and filling of the grains were also better than local varieties. No rust disease was observed. The yield from the experimental plot was 1500 kilograms per acre. We arranged a demonstration and invited farmers from Konkan. They were greatly impressed to see the performance of the Mexican variety of wheat. They decided to introduce this variety during the next winter season.

Accordingly, during the winter of 1967, the Thane Zilla Parishad secured large quantities of Sonora-64 and Kalyan Sona for distribution among farmers. Both these varieties performed well. The Konkan farmers who were growing only one crop of rice for generations, introduced wheat as the second crop. The wheat crop has now come to stay in Konkan.

Dr. Norman Borlaug made a great contribution in development of Mexican varieties of wheat for which he was awarded Nobel peace prize. He was invited by Government of India for felicitation. I went to Delhi to meet him. He was observing experimental plots at Indian Agricultural Research Institute, New Delhi. Dr. Swaminathan introduced him to me. I narrated our experience of the introduction of wheat in Konkan region, which was a traditional rice growing area. I further told him that Mexican varieties which he evolved, were greatly responsible for its success. However, he immediately replied that it was not the variety alone but the efforts of farmers and scientists which were more important. This shows the humility of a great scientist who revolutionised food production in the world.

The felicitation function of Dr. Borlaug was organized at Vigyan Bhavan, New Delhi. The President of India presided over the function. Dr. Borlaug said, "I found poverty and hunger in many parts of the world. Millions of children go to sleep at night without taking a meal in the evening. I had therefore decided to utilize my knowledge to increase wheat production. I left U.S.A. and went to Mexico. I tried to do research on development of early dwarf varieties which would also be resistant to diseases. In trials they gave higher yields than local varieties. Scientists from all over the world were invited to see the performance of these varieties. India was one of the first countries to experiment with these seeds. After seeing dramatic yields of these varieties, India decided to import large quantities of Mexican wheat seeds. Accordingly, eighteen thousand tonnes of seeds of Mexican varieties were shipped from Mexico to India. There is no precedence in the world of shipment of such large quantities of seeds from one country to another".

"The Mexican varieties of wheat became very popular among the farmers of Punjab, Haryana and Western Uttar Pradesh. These varieties gave double or triple yields than local varieties and made a revolution in wheat production which is popularly known as 'Green Revolution'. I give great compliments to farmers, scientists and administrators for their contribution. I also congratulate the Government of India for evolving public policies suitable for food production. I am convinced that if we want to eliminate hunger on the earth we have to adopt high yielding and hybrid varieties of food crops". The speech of Dr. Norman Borlaug greatly inspired me to continue my efforts for introduction of wheat crop in Konkan.

7

HORTICULTURE

Prosperity through Chikoo Nursery

The Northern part of Thane district is known for its horticulture. The prosperity in this area was created mainly by the Sapota (Chikoo) fruit tree. This fruit tree was introduced in this area in the early part of this century. However, it could not spread for 50 years. One of the reasons for this was lack of nurseries. Two Irani fruit growers were preparing some chikoo grafts but they were available to only a few farmers. Farmers in this area therefore did not think of cultivation of chikooes.

The Kosbad Institute was established in 1949. Two main programmes were started in this Institute. The first was to impart skill training to the sons of farmers and the second was to supply quality seed and plants to farmers. We started a horticultural nursery for supply of grafts of chikooes. This was the first horticultural nursery started by a voluntary organization. Shri Vasudeorao Chitre was the first director of this nursery.

Large number of farmers were visiting the Institute to see demonstrations of rice crops. We started showing them the technology of preparing grafts of chikooes. We also advised them to plant chikoo trees alongside rice crops. This would create a new farming system of food and fruit crops. The farmers were convinced of this approach and they carried grafts of chikooes with them. In this way the chikoo tree spread all over Thane district. We studied the growth of this fruit tree in different agro-climatic conditions of the district. The farmers from Maharashtra, Gujarat and Karnataka also started coming to take chikoo grafts.

After the establishment of cooperative sugar factories in Maharashtra the sugarcane growing farmers purchased chikoo grafts in large numbers for planting on their lands. The Agricultural Department of Maharashtra Government and the Agricultural Universities in Maharashtra and Karnataka secured chikoo grafts from this nursery and planted them as mother plants in their horticultural nurseries. The chikoo grafts which were prepared here were of "Kalipatti Variety". This variety was of very good quality and one of the best in the country. We were therefore able to introduce the best variety of chikooes among the farmers through our horticultural nursery at Kosbad Hill.

The greatest advantage of this nursery was available to the farmers from the northern part of Thane district. The farmers from this area were growing vegetables after the harvest of their rice crop. However the prices of vegetables crashed during this season and farmers were put to great loss. On the other hand the prices of chikoo fruits did not come down. We arranged farmers' rallies and explained to them the importance of planting chikoo trees. The farmers responded positively to our appeal. Quality grafts of chikooes were made available to them from the nursery of the Agricultural Institute. Every farmer having an irrigation well started planting chikoo grafts on his land. Even small and marginal farmers took up chikoo plantations. They adopted a mixed farming system of chikooes, rice and vegetables. The chikoo tree requires about five years for fruiting. During this period the farmers were getting income through rice and vegetables as intercrops. Some farmers grew floricultural crops like lillies and tuberoses as intercrops with chikooes.

Mixed farming of chikooes, rice, vegetables and flowers generated gainful employment in the area. The tribals and landless labourers benefited from this. The small and marginal farmers got cash income through sale of chikoo fruit, vegetables and flowers. Their purchasing power improved. The chikooes from this area are sent to various markets all over India. Considerable amount of employment was also generated in the transport sector. Small Workshops have been started for repairs of tractors, power tillers, electric pumps and such other agricultural machinery. The local mechanics got employment due to such workshops. Chikoo grafts require pots on a large scale and so the local potters too benefitted. Similarly, large number or bamboo baskets are required for packing fruits. The landless basket weavers got employment for preparing baskets. In this way gainful employment was generated and local skills were revitalized by the chikoo plantations.

As more and more farmers took to planting chikooes, the demand for grafts increased. Over one million chikoo grafts were prepared in the nursery of the Kosbad Institute during the last 40 years. This has enabled the farmers to undertake chikoo plantations on an area of about ten thousand hectares. It is noteworthy that most of these plantations were done by small and marginal farmers. This is a unique example of how a horticulture nursery brings about prosperity in an area.

Technology for Preparation of Rayan Seedlings

The horticultural nursery at Kosbad Institute was started in 1949. Mango and chikoo grafts were prepared in this nursery. Seedlings of lemon were also prepared. However, the emphasis was on preparing chikoo grafts. The region had large area under 'Kalipatti' variety of chikoos. The leaves of this trees are broad and remain evergreen. The pulp of the fruits in this variety is soft while in other varieties it is granular. 'Kalapatti' variety has therefore become popular due to its good quality fruits.

Formerly chikoos were propogated by air layering method. But now they are propogated by grafting on Rayan stocks. The layers have fibrous roots and could be grown only in sandy soils. Such soils existed near the coastal tracts only. Most of the soils in the area are clayey. If the layers are planted in clayey soils their leaves become yellow and later fall down on the ground. The trees become unhealthy in clayey soils. During cyclonic weather the trees fall down on the ground due to weak roots.

Contrary to this the chikoo grafts prepared on rayan stocks are very hardy due to tap roots. Such grafts are suitable for growing in clayey soils. During cyclonic weather the trees remain erect and do not fall down on the ground. They have moderate tolerance for brackish water and they are also drought tolerant. The preparation of chikoo grafts on rayan stocks was developed in the early Forties. Since that time there has been a fast growth in cultivation of chikooes.

The main constraint for preparing chikoo grafts was the scarcity of rayan seedlings. We were collecting these seedlings from the villages of Wadhvan and Varor. In these two villages large number of rayan trees existed. The seedlings were available around these trees. The landless labourers from these villages uprooted these seedlings very crudely which resulted into damage of roots. The mortality of such seedlings was very high. However, we had to continue collecting the seedlings from these two villages for several years. Thereafter we started bringing rayan seedlings from 24-Parganas districts of West Bengal. There are a large number of nurseries who prepare rayan seedlings in that area. Due to alluvial soils the uprooting of the seedlings was comparatively easier and the root system was not damaged. These seedlings were transported by railways and it took more than a week in railway transit. The practice was to prepare a mud ball around the roots which helped to keep the seedlings live. Mortality of seedlings collected from 24-Parganas district was less as compared to seedlings collected from Varor and Wadhwan villages.

During 1963 we conducted an experiment to prepare the rayan seedlings on the farm of the Kosbad Institute. We had sown the seeds of rayan on raised beds. However, there was no germination of seeds. In 1966 we conducted another experiment to give heat treatment to the seeds before sowing, using an egg incubator. This incubator was run on Kerosene, maintaining a temperature of 40 degree celcius. The seeds were kept for 24, 48, 60, 72, 96, 168 and 264 hours. To maintain humidity levels the seeds were kept in wet cloth. We found that the seed which was kept for 24 hours at 40 degree celcius had germination of 48 per cent. I wrote an article on this experiment which was published in the April 1974 issue of *Indian Journal of Agricultural Science*. The title of the article was "Note on Activation of Germination of the Seeds of Spanish Cherry". This experiment proved to be very useful for preparing rayan seedlings. We became self reliant in production of rayan seedlings, which enabled us to expand the programme of grafting on chikooes.

The young farmers in the area who had horticultural nurseries were bringing rayan seeds for treatment at Kosbad Institute. We placed Community incubators for treatment of rayan seeds. Thereafter, we had developed a simple technique of treatment of rayan seeds. The seeds were soaked in water for 24 hours. The treated seeds were then kept in small gunny bags and the bags were placed in a pit of cattle manure for 24 hours. There was considerable heat in the pit due to decomposition of cattle dung. This helped to activate the rayan seeds. We observed that seeds treated in this way had more than 70 per cent germination rate. This technique of heat treatment to rayan seeds through manure pit has become popular due to its low cost. It has enabled the nursery farmers to become self reliant in production of rayan seedlings, which are used as stock plant for grafting chikooes.

New Method of Preparing Papaya Seedlings

In the northern part of Thane district farmers grow papaya as intercrop in chikoo plantations. The farmers used to prepare raised beds and sow the papaya seeds in the first week of May. The seedlings were uprooted from the raised beds in the middle of June and then transplanted in the fields. Due to uprooting, the roots were being disturbed. Many seedlings would suffer and there was a high rate of mortality. The seedlings which survived were also weak and did not have good growth. We found that the roots of the papaya seedlings were very tender. If these roots are disturbed, it adversely affects the growth of plants and it takes longer time for fruiting. Such plants bear fruits in the summer season when there is no demand for papaya.

To avoid disturbing the roots, we prepared seedlings in polythene bags. However, the leaves of such seedlings remained yellow. Some roots came out of polythene bags. We saw that polythene bags were also not suitable for raising papaya seedlings. We planted these seedlings in middle of June. There was no mortality in the transplanted seedlings. However, the growth of the plants was not so healthy. These plants got the fruits in March, about two months earlier than seedlings uprooted directly from raised beds. The fruits received good market rates.

In another experiment, the seedlings were prepared in small bamboo baskets. Dried leaves of the *Palas tree (Butea monosperma)* were placed at bottom of the basket. Sieved soil mixed with compost was filled in the basket. Four seeds were sown in each basket in the first week of May. The baskets were kept in partial shade. The seeds germinated after two weeks. Two healthy seedlings were kept in each basket and others were uprooted. The growth of seedlings raised by this method was very fast. The seedlings were planted along with bamboo baskets (which are bio-degradable) in the permanent field in second week of June. By this time the rains had also started. The bamboo baskets rotted after some time. The leaves of these plants remained evergreen and the growth was very vigorous. The plants flowered in middle of August and fruiting started in middle of December. We sent these fruits to the Bombay market and fetched very good price due to the Christmas season. It was winter season and there was good demand for papaya fruits in Bombay market.

We found that preparation of seedlings of papaya in bamboo baskets helped to keep their roots undisturbed which enabled the plants to have early bearing of fruits. I wrote an article on this experiment viz., "Papaya can surely be more paying", which was published in *Indian Horticulture* magazine of July-September 1972.

We arranged one demonstration of preparation of papaya sedlings in bamboo baskets which was attended by a large number of farmers. The farmers have now adopted this practice and are getting early harvests from their papaya plantations. There are large number of landless artisans who prepare bamboo baskets in this area. Due to increase in demand of small bamboo baskets for papaya nurseries, employment was also generated for landless artisans.

Natural Farming

A plantation of sapota (chikooes) had been established on five hectares of land on the farm of Agro Industries and Tribal Welfare Foundation at Manpada. We followed the 'Natural Farming' method in this plantation since 1982. "No Tillage" practice was adopted in the method. Ploughing or tilling was not done by any implement, but the work of tilling the soil was done by earthworms. We found that the earthworm population in the soil was getting destroyed due to continuous tillage and excessive use of chemical fertilisers and pesticides. We had therefore refrained from the use of these three practices. Thrust on the other hand, was given on preparation of organic manures. All the grasses and weeds were utilized for preparing compost. Leaves of subabhool are also used for this purpose. The compost was spread on the soil around the sapota trees. A thin layer of dry grass or dried leaves was spread on the ground, which served as mulch and helped to retain moisture in the soil. Moist condition is congenial for earthworm activity, and dry grass or leaves gives shelter to earthworms and protects them from attack of rodents or birds.

We introduced leguminous fodder of *Stylosanthes hamata* which helped to fix nitrogen in the soil thereby increasing its fertility.

Shallow channels of half meter wide and half meter deep were laid between two lines of sapota trees. By adopting this method less water was required as compared to the traditional basin method. The animal manure and oil cakes were applied in the channels. All the feeding roots of sapota are found in these channels. Similarly due to moist conditions the earth worms use channels as shelter during the summer months. These channels serve as drains during rainy season.

Chemical fertilizers or pesticides were not applied to the sapota trees for the last fifteen years. The trees take their nutrients from organic manures and manure prepared by earth worms. The trees develop natural resistance and remain free from pests and diseases. We found that by following the natural farming method, productivity of sapotas improved considerably. The quality of fruits both in colour and taste too improved.

Shri Bhaskar Save, a progressive farmer in Deheri village in Umbergaon Taluka of Valsad District in Gujarat State has been following the 'natural farming' method on coconuts for several years. He has also adopted 'no tillage' practice and applies only organic manures to his soils. These two conditions are congenial for the growth of earthworms. With the adoption of natural farming the productivity of coconuts increased greatly. His farm is being visited from people all over the world.

The farmers in Thane district who have sapota and coconut plantations have also adopted the natural farming method. I found that through the "natural farming" method, agricultural development could be achieved together with environmental balance.

Creation of Seperate Department of Horticulture

Shri A.R. Antulay, the Chief Minister of Maharashtra visited the Kosbad Institute on 4th April, 1981. We took him to Chimbave village to show the farm of Shri Udharya Bujad, a tribal farmer. The area of his farm comprised only two acres. He constructed a small well and installed a pump. He grew vegetables like tomatoes, onions and bottle-gourd. He also planted fifty trees of 'Sapota'. Besides, fruit trees like mangoes, custard apple, jackfruit and *ber* were grown on his farm. He established a small nursery for preparing mango, sapota and coconut seedlings. He also kept three bee boxes. A bio-gas plant was installed which supplied cooking gas. His house was prepared from bamboos and the roof was made of Managlore tiles. It was a clean and neat dwelling.

Shri Bujad told the Chief Minister that he was growing only one crop of rice between June to September. This crop was often subjected to vagaries of monsoons and the yields were very low. He could not support his family from the rice produce throughout the year. After the harvest of rice in October, he did not have any employment on his land. He and his family members had to roam about in search of labour either in the forest areas or in neighbouring horticultural gardens. However, the opportunities for such employment were very meagre. Many a times, he had to spend days without any work.

Shri Bujad told the Chief Minister that in 1967-68 the Kosbad Institute helped him to construct a well and install the pump. The Institute also supplied grafts of fruit trees and vegetable seedlings for planting. Due to well irrigation his rice crop had become assured. He had now adopted high yielding varieties of rice which enabled him to become self-reliant in food production.

Shrimati Bujad told the Chief Minister that everyday she carried her vegetables for sale in the nearby markets, which earned her about Rs. 30 to 40 per day. Besides, many tribals from the surrounding areas also visited their garden to purchase vegetables. While formerly she had to go to market to purchase vegetables, now she went there to sell them.

Shri Bujad told the Chief Minister that he was sending his sapotas to Crawford Market in Bombay. He packed the fruits in bamboo baskets, which were carried by the trucks to Bombay in the evening. He also sold other fruits in the nearby markets. The seedlings of mango, sapota and coconuts were sold to nearby tribal farmers. Shri Bujad told the Chief Minister that he was getting an annual income of about Rs. 20,000 from his two acres of land. More importantly, the five members of his family got gainful employment on his land throughout the year.

After seeing the garden of Shri Bujad, Shri Antulay told me "I am convinced that it is through horticulture we could give stability to small farmers and tribals. I have, therefore, decided to announce creating a new Department of Horticulture in Maharashtra State. For that purpose, I will require your help." I assured him that I would extend all possible help. On hearing this, he told me, "I do not simply require your help but you will have to accept the responsibility of this department." However, I explained to him that I have commited myself for the tribal uplift work for the last thirty years and I am unable to leave Kosbad Institute immediately. Shri Antulay agreed with my views and suggested me to accept the responsibility of Honorary Adviser to Government of Maharashtra in the Department of Horticulture. I agreed to his suggestion.

Thereafter we went to Kosbad Institute. An exhibition of fruits grown by the local farmers had been arranged. They included mangoes, sapotas, pomeloes, litchies, pomegranates, *Kagdi* lime, guavas, coconuts and arecanuts. After seeing this exhibition Shri Antulay was convinced about the importance of horticulture in the development of small farmers and tribals.

We arranged a farmers rally during the visit of Shri Antulay and about 5,000 farmers from all over the Konkan attended the rally. While addressing, the rally Shri Antulay said,

> "I visited a garden of a tribal farmer, who had only two acres of land. He planted fruit trees like sapotas, mangoes and coconuts and grew variety of vegetables as inter-crops. I found that a small tribal farmer has achieved stability in his life due to adoption of horticulture. I am, therefore, convinced about the importance of horticulture in the upliftment of small farmers and tribals. I, therefore, make an announcement that a new Department of Horticulture will be started in the Government of Maharashtra. This department will have a Minister, Secretary and Director. Dr. Jayant Patil will be appointed as the Honorary Adviser to the Department of Horticulture".

There was wide appreciation from all farmers of the Maharashtra State for setting up a new Department of Horticulture in the Government of Maharashtra. The farmers of Konkan were pleased as the region was hilly and there existed a great potential for growing two important fruit trees namely, mangoes and cashews. Both these fruits were in great demand both in the domestic market and for export. Similarly, the farmers from dry areas of Maharashtra were glad as the drought resistant fruit trees like *ber*, custard apple and mangoes could be grown in this region.

Maharashtra had 78 per cent of its area under rainfed or dry conditions and, hence dry land horticulture would bring about economic development of small farmers in these areas. The decision of the Chief Minister, Shri A.R. Antulay for the creation of a seperate department for horticulture was a step in the right direction.

Chief Minister, Shri Antulay called me to Bombay and alloted a chamber in Mantralaya for setting up my office. I was also provided the necessary staff. I undertook an extensive tour of Maharashtra and studied the present condition of horticulture in the State. I found that the farmers could not adopt horticulture due to lack of planting material. Similarly, there were no bankable schemes for financing horticulture. We, therefore, took up the first programme of establishing horticulture nurseries in every district. For that purpose, the seed farms of Agriculture Department were transferred to the newly created Department of Horticulture. Good quality planting material was secured from the Agricultural Universities and Departmental Farms from Maharashtra and other fruit growing States. These planting materials were utilized as mother plants for preparing grafts and plants. A central nursery was also established at Dapchary in Thane District for preparing mangoes and sapota plants. Due to establishment of these nurseries, quality planting material was made available to the farmers. I found that for horticulture development nurseries played a crucial role.

A meeting was organized with the representatives of Nationalised and Cooperative Banks for preparing crop-wise bankable schemes. This was of considerable aid in providing financial support to undertake the plantation of horticulture crops. I found that crop-wise bankable schemes were necessary for adequate implementation of horticulture development.

Due to the establishment of the nurseries by the horticultural department quality planting material became available to the farmers. Similarly due to financial support from the nationalised and cooperative banks the farmers were now able to undertake horticultural plantations. The Government of Maharashtra strengthened the horticultural development programme by linking it with its Employment Guarantee Scheme from May 1990. Maharashtra has now emerged as one of the leading States in the country in terms of horticultural development. The fruits from Maharashtra are being marketed all over the country, besides being exported to various other countries. Horticultural development in Maharashtra has played a pivotal role in enabling the small and marginal farmers of the State to attain economic stability.

8

TREE PLANTING AND WASTELAND DEVELOPMENT

Banyan Tree

There was a big banyan tree *(Ficus benghalensis)* in front of the Kosbad Institute. One could watch the whole valley for about five Kilometers and the Arabian Sea by sitting under this tree. Since the tree was circular in shape, there was shade throughout the day. A number of meetings were held under this tree. Mrs. Godavari Parulekar, a marxist communist leader, who was working among the tribals during the forties, used to organise meetings under this tree. The tribals in this area revolted against exploitation by the landlords, moneylenders and forest contractors in 1945 which is known as the "Warli revolt." Every tribal person coming for the meeting brought one stone from his village and a heap of stones was arranged under this tree to show the strength of the people. When the Kosbad Institute was started in 1949, we prepared a platform under this tree. The people who were passing by on the road used to sit under the shade of this tree in the afternoon. We constructed a small cement tank to provide drinking water.

We were holding the classes for students under this tree. It reminded us of the atmosphere of the "Shanti Niketan" of Gurudev Rabindranath Tagore. The foundation stone of the Agricultural School was laid in 1950 at the hands of Shri R.K. Patil, who was then the Food Commissioner, Government of India. At that time, the meeting was held under this banyan tree. The inauguration of the Agricultural School was done at the hands of Shri Balasaheb Kher, Chief Minister of Bombay State in 1951. We arranged a big meeting at that time under this banyan tree. All the meetings of farmers were also held under this tree.

In March 1957 at about 10 O'clock in the morning we heard a big sound and to our surprise, found that the whole banyan tree was uprooted and had fallen on the ground. The tree which gave shelter to thousands of passers-by and where historical meetings were held, was no more.

When we observed the fallen tree we found the whole trunk had become hollow. Besides, there were no aerial roots. It was therefore, not possible for the trunk to bear the heavy weight of the crown of the tree. We then cut the branches of this tree and cleared the site.

We decided to plant a new banyan tree, at the same place where the old tree existed. For that purpose we dug a big pit and filled it with compost. Two healthy cuttings from branches of a nearby Banyan tree were prepared. These cuttings were planted in the second week of June when the rains started. A cactus fence was erected to protect the tree from stray animals. The cuttings started sprouting within a month and they were full of leaves during the rainy season. We provided hand watering to this new tree during the summer. The watering was done only for one year.

Thereafter it did not require any watering. Within ten years, it grew up into a big banyan tree and it looked like the old tree which existed before. Swami Anand told us the history and greatness of the banyan tree. He said that this tree lives for hundreds of years and keeps its surrounding micro climate cool. The tree gets fruits twice or thrice a year which attracts birds. Due to singing of birds the whole atmosphere remain pleasant. He advised us to plant more banyan trees.

Swami Anand narrated from an ancient scripture about the greatness of a tree :

> "Ten wells are equal to one tank.
> Ten tanks are equal to one lake.
> Ten lakes are equal to one son.
> Ten sons are equal to one Tree."

We recieved great inspiration from this scripture and planted number of banyan trees on roadside of Kosbad Hill.

Planting Neem Trees

Swami Anand used to advise us that we should plant neem trees on the campus of the Kosbad Institute. He told us that there is no other tree like neem for purifying the air. He told us a good story about planting of neem trees. Mahatma Gandhi was staying at Sabarmati Ashram near Ahmedabad. Shri Totaram a young man from Fiji Islands, came to stay in the Ashram. Every inmate of the Ashram had to do some manual labour. Totaram asked

Gandhiji what work should he do? Gandhiji replied that he should take up the work of planting neem trees, as the whole area was barren. Accordingly Shri Totaram started planting neem trees. He went to surrounding villages where there were neem trees. He collected the seedlings and brought them to Sabarmati Ashram. He constructed pits and planted these seedlings very carefully. He used to bring water in buckets from the river bed of Sabarmati as there was no tap water in those days. It was a very hard work but he accepted it as a mission. During his entire life he only did the work of planting neem trees. Due to his efforts the area from Sabarmati Ashram to Sabarmati Prison was covered with neem trees. The atmosphere there in summer became very cool and pleasant. Swami Anand used to say that it was the contribution of Shri Totaram that made the Sabarmati Ashram bloom with neem trees.

We started planting neem trees in the campus of the Kosbad Institute. For that purpose, we prepared small pits in summer. The pits were filled with compost and good soil. When the rain started in the second week of June, we planted sprouted neem seeds in these pits. Such seeds are available under the old neem trees. We prepared cactus fence to protect the seedlings from stray animals. When the seedlings were two months old, one healthy seedling was kept in each pit. After the rains were over in October, we started hand-watering these seedlings. This work was done by the students of the Ashram schools and secondary schools. Each student was given the responsibility to look after one seedling. The tribal students took great care of the seedlings. Within one year the seedlings grew very fast. We found that the neem is a very hardy plant and it requires hand-watering for only one year.

The whole campus of the Kosbad Institute, Ashram schools and secondary schools bloomed with neem trees. Every morning, birds sit on the trees and sing. During the summer too the neem trees remained green thus leading to a pleasant atmosphere.

Neem has multiple uses. The seeds produce non-edible oil used for making soap and other products. Pesticides prepared from Neem are considered to be eco-friendly. Oil cakes produced from neem are used as manure due to their insecticidal property, the crops thus remain free from insect attacks. I learned that a German Scientist by the name of Dr. Smuterer while moving in the forests of Sudan found that all the trees in the forest were prone to locusts. Neem was a sole exception to this. He therefore undertook research relating to this plant. Many scientists who werè inspired by his research too conducted experiments on this plant. Presently scientists from U.S.A., Japan, Germany, U.K. and many other countries are devoting themselves to research on neem.

Research has revealed that if the neem cake is blended with urea fertilizer, leaching does not take place. This is very important for crops like rice. The green leaves of neem could be used as green manure for rice crop, while the dry ones could be used as compost. The green leaves could also be used for preserving seeds and grains. Often small branches with green leaves are tied in houses to prevent and fight against certain diseases like measles. This reveals that neem leaves inherently have antiseptic properties. The tender branches of neem are also used for cleaning teeth by the rural people. Tooth-paste too is prepared from neem. In the State of Maharashtra, on *Gudi Padava*, the New Years Day, people consume neem leaves to keep away parasites in the intenstines.

Neem is a very hardy tree. It grows in dry and draught-prone areas. It gives seeds in the month of April and May. Landless labourers are employed for collecting the seeds. Oil is extracted by artisans through oil ghanis. The neem is thus a very important tree for generating gainful employment in dry areas. We prepared a nursery of neem seedlings at the Kosbad Institute by involving school children from the surrounding villages. The children celebrated Tree Planting Day by planting neem trees and have contributed to greening all the villages.

The Waki Project

One day in the morning Shri Gokhale, Collector of Thane came to Kosbad Institute without prior intimation. He was accompanied by a team of revenue officers. He told me that he was camping at Dahanu and enquired whether any Government land was available near Kosbad. He was told by his officers that a piece of land from woodland protected forest was available at Waki which was about three kilometres from Kosbad. He, therefore, made a visit to the Institute after inspection of this land in the morning. He told me that he proposed to recommend to the Government to grant this land to Kosbad Institute. Shri Gokhale served as Deputy Secretary in Rural Development Department prior to his appointment as Collector of Thane. He knew the tribal development work we were doing at the Institute. He, therefore, felt that the land at Waki would be useful for tribal development projects.

Shri Gokhale took us to the site at Waki. The land was degraded due to overgrazing and soil erosion. No vegetation existed on the land except some thorny bushes. Since the land was situated on both the sides of Kosbad-Dahanu Road and had easy access we agreed to the proposal of Shri Gokhale. We were confident that with our experience at Kosbad Institute, we could develop this degraded land. We thanked Shri Gokhale for his gesture.

Shri Gokhale recommended the proposal of this land to the Commissioner of Division. I met Shri Janardan Patwardhan who was Commissioner at that time. He told me that he was informed about the tribal development work done at Kosbad Institute. However, he had not been able to visit the Institute till then. He was, therefore, keen to visit the Institute. He further told me that he would bring his brother Shri Achyut Patwardhan with him to show the tribal development work at the Institute.

After a week Shri Janardan Patwardhan and Shri Achyut Patwardhan came to Kosbad Institute. I knew Shri Achyut Patwardhan as a great national leader in the freedom struggle of India. His role as a revolutionary in Quit India Movement had given great inspiration to me during my student days. Shri Patwardhan later became a disciple of J.Krishnamurti and followed the philosophy of his *Guru* (teacher). We were very happy to receive Shri Achyut Patwardhan at Kosbad Institute.

We showed him various tribal development activities like Ashram School and Agricultural Training Centre for tribal youths. We took him to surrounding villages to show the project of small irrigation wells which enabled the tribal farmer to grow vegetables and fruit trees thereby getting gainful employment on their lands. After seeing this project he remarked "This is a live example of development with social justice. The tribals who worked as labourers for generations now got the opportunity to develop their own vegetable and fruit gardens. This would enable them to lead life with self respect." He complimented the staff of the Kosbad Institute and desired that the project of small irrigation wells be extended further. We were greatly inspired by the visit of Shri Achyut Patwardhan.

Alongwith Shri Janardhan Patwardhan we went to the site of the land at Waki. We told him that we proposed to utilize this land for development of seed farm and horticultural nursery. The seeds and plants would be distributed to the tribals in Thane District. Shri Patwardhan appreciated our idea and recommended to the Revenue Department of Maharashtra Government to grant this land to Kosbad Institute.

The Revenue Department raised several queries about this proposal. I personally had to visit Bombay several times to meet the concerned officers in Sachivalaya. This helped to expedite the proposal. Shri Moreshwar Thakur who was working as Deputy Secretary in Revenue Department rendered me valuable help in this matter.

One day we received a letter from the Revenue Department that I should come to Bombay and meet Shri Balasaheb Desai, Minister for Revenue. Accordingly, Acharya Bhise and myself met Shri Balasaheb

Desai in Sachivalaya at Bombay. Shri Pande, Joint Secretary, Revenue Department was present with the concerned file. Shri Desai asked me the first question about utilization of this land. I told him that we proposed to set up a seed production farm and horticultural nursery on this land. We also proposed to distribute the seeds and plants to the poor tribals from Thane district. An Ashram School for tribal children would also be set up where training in agriculture would be imparted through work experience projects. Shri Balasaheb Desai knew about the tribal welfare work at Kosbad Institute. Besides, he knew me personally as Shri Desai and my father had worked together as General Secretaries of Maharashtra Pradesh Congress Committee for four years. He was also a colleague of my father for five years in the Cabinet of Shri Yashwantrao Chavan, Chief Minister of Maharashtra.

Shri Balasaheb Desai told the Joint Secretary Shri Pande, "Acharya Bhise has devoted his whole life for upliftment of the tribals. I consider him to be my *Guru* (teacher). Shri Jayant Patil works like a missionary among the tribals. Even though he received higher education in United States he came back to India to work among tribals. I have seen the work of the Kosbad Institute and we must consider the proposal for granting of land to the Institute."

Shri Pande explained to the Minister that out of 35 acres of Government land the Commissioner, Konkan Division has recommended 20 acres to be granted to the Institute. At this point Shri Balasaheb Desai told us "even though, the Commissioner has recommended 20 acres of land I grant 35 acres to you. I am confident that you will make the best use of this land for the welfare of tribals." This shows the trust which the Honourable Minister had in voluntary workers like us. Acharya Bhise and I thanked Shri Balasaheb Desai for this noble gesture and promised him to make the best use of the land.

Shri L.W. Choudhari who was then the Deputy Collector of Dahanu handed over the possession of 35 acres of land to us. Shri Chaudhari did very good work of tribal welfare in Thane district. He was very happy to know that this land would be utilized for educational and agricultural projects of tribals.

We prepared a project of horticultural nursery which included land development, irrigation well with pump, pipeline and planting material. We submitted it to 'People's Action for Development' a donor agency which was working under the Ministry of Agriculture, Government of India. Shri J.B. Singh who worked as Joint Commissioner in this organization visited the Kosbad Institute. Shri Singh had considerable experience of working with the non-government organizations. We took him to Waki and showed him the site where the project of horticultural nursery was to be set up. We also showed him the horticultural development work which we were doing among tribals. On his return to Delhi, he placed our proposal before the Governing Council of the organisation, which gave its approval to this project. We received encouragement and support from People's Action For Development and particularly Shri J.B. Singh in setting up this project.

We undertook first programme of land development on ten acres. The land was slopy and due to soil erosion large gullies were formed. We, therefore, prepared terraces according to slopes. We then constructed a well and installed pump for irrigation. We planted mangoes, cashews, custard apple, *ber,* chikooes, coconut and guavas as mother trees for preparing quality grafts and plants. Papayas, bananas and vegetables were grown as intercrops. We held demonstrations which were attended by large number of tribal women. We told them that this nursery had been set up to supply vegetable and fruit plants for planting on their lands. The objective was to overcome the problem of malnutrition through production and consumption of protective foods consisting of vegetables and fruits. At the end of the demonstrations every tribal women carried seedlings of vegetables and fruit plants like papaya and bananas. They set up Nutrition Gardens' in their backyards. After two years grafts of fruit trees were also supplied to tribals which enabled them to establish small fruit gardens. In this way the horticultural nursery proved to be an important instrument for overcoming problem of malnutrition among tribals.

The Community Aid Abroad from Australia gave support for development of remaining twenty five acres of land. It consisted of construction of another well and setting up of pump and pipeline, construction of a seed store and purchase of a power tiller. We utilized this land for setting up seed farms of rice and pulses (grain legumes) which were the main food crops of the tribals. Seeds of high yielding varieties of rice and improved varieties of pulses were produced for distribution. This had improved the productivity of rice and pulse crops of tribals.

Some tribal women whose land holdings were very small told me that even though they could obtain good rice seeds from our seed farms they were unable to raise rice nursery in time due to lack of implements like ploughs. They were unable to maintain a pair of bullocks required for ploughing as they did not have sufficient fodder. They had to depend on ploughs of neighbouring farmers which were available only after their work was over. This resulted into late sowing which adversely affected their rice yields.

From the next year we prepared a community rice nursery on an area of about 5 acres. We found that there was a great demand for these seedlings. The tribal women who were unable to prepare their rice nurseries due to lack of inputs, benefited by this project. This project became very popular and so it was extended on an area of about 10 acres. We found that community nursery was one of the very effective programmes for introduction of high yielding varieties of rice among the very small land holders.

The Minister for Education, Shri Madhukarrao Chaudhuri visited Kosbad Institute. We took him to Waki to show the horticultural nursery. He was greatly impressed to see the site. He felt that this site was suitable for starting an Ashram School. We told him that if an Ashram School is started it will benefit the tribal children from Waki, Kankardi, Aagar and Narpad villages. Shri Chaudhuri desired that we should send the proposal of the Ashram School immediately. Accordingly he sanctioned one Ashram School at Waki.

The Ashram School at Waki had a devoted staff. Smt. Kalawati Pardhi who was Headmistress of this school belonged to scheduled tribe community. She received her education from the secondary school at Kosbad and so she developed idealism in her life. Her husband Shri Sakharam Pardhi took training in agriculture at the Kosbad Institute. He was a very committed teacher. Due to their devotion the Ashram School at Waki became an ideal Institute. Besides, giving academic education to the students of the Ashram School we gave them training in agriculture, horticulture through the work experience projects. This has enabled them to learn skills in agriculture. The students while going to their homes in vacations carried samples of seeds and plants. This has enabled them to apply their skills in high yielding varieties of rice, vegetables and fruit trees.

We feel that if the wastelands are properly developed and projects like community nursery of rice and horticultural nurseries are established, it brings about agricultural development of the weaker sections. We also observed that if educational institutions like Ashram Schools are attached to these projects it enables to impart skill training in agriculture. Such skill training was found to be very useful in generation of self employment which is the need of the day.

Development of Wasteland at Manpada

Keeping in view the need to disseminate technology among the tribal farmers of Thane district we established the Agro Industries and Tribal

Welfare Foundation in 1974. Shri Vasantrao Naik, Chief Minister of Maharashtra inaugurated the Foundation. I was assigned to take over as its Managing Trustee.

The Government of Maharashtra granted twenty hectares of land to the Foundation to set up a seed farm and horticultural nursery at Manapada in Jamboogaon village of Dahanu Taluka in Thane district in 1978. This was a wasteland and was highly degraded due to soil erosion and over grazing, which led to the formation of large gullies. Not a blade of grass grew on this land, which was simply used for walking cattle.

We prepared a project for the development of this wasteland. The Community Aid Abroad (Australia) and People's Action for Development (Maharashtra) agreed to give financial support to this project. We first laid out a bund of one meter height and one meter breadth around the twenty hectares in summer. As soon as rains started we planted four lines of cactus on this bund. The cactus was collected from the nearby villages and was used to protect the proposed farm from the menace of stray animals. In the first year itself we observed growth of grass, since the land had been protected from stray animals. This shows that if a simple cactus fencing is done on the wastelands it helps in the growth of the grasses which were otherwise dormant due to overgrazing by animals.

After the rains were over in October we undertook the programme of terracing. For that purpose a survey was made and the terraces were prepared according to contours. With a view to protect the soils from erosion a pitching of stones and rubbles was done on the border of the terraces.

We constructed four dug wells for irrigation. Pumping sets were installed for lifting the water. A galvanised pipeline was laid for irrigation. We prepared a grid of this pipeline connecting all the four wells. We laid out necessary roads and built culvarts on the streams passing through the farm.

Since the land was barren for years it was lacking in organic matter. We therefore, replenished it with organic manures like cattle manure and compost. In the first year we grew 'sunhemp' as a green manure for further replenishment of this barren land. We found that organic manures were essential for development of wastelands.

We undertook a programme of fodder development on this wasteland, and for this purpose we prepared a nursery of *Stylosanthes hamata* in summer. The seedlings were transplanted in the second week of June when the rains started. The *Stylosanthes hamata* grew very well on this land. We found that *Stylosanthes hamata* is one of the suitable fodders in the wasteland development programme. We planted seedlings of subabhool on both the sides of the roads. Leaves of subabhool could be used as fodder. Tribals in the area collected the leaves of subabhool for feeding their cattle. Subabhool is one of the most suitable fodder tree in the wasteland development. Subabhool could also be used as a fuel wood tree, stems and branches are suitable for that purpose. We planted casurina which is a fast growing fuelwood tree. This tree also grows fast under the agro-climatic conditions of Konkan. We planted *Shivan (Gmelina arborea)*, teakwood and rosewood. These trees could be used for their timber. Bamboos too were planted, which were in great demand for preparing baskets for packing vegetables and fruits.

We planted chikoo trees on this land. We found that due to high calcium content in the soil the growth of the plants was very good. Mangoes, litchis, pomeloes, coconuts, cashews were the other fruit trees planted on this land. Waste lands if properly developed could be utilized for establishing fruit orchards.

A nursery of chikooes was set up, where the chikoo grafts were prepared on rayan seedlings. The grafts were distributed to the surrounding tribals. We prepared mango grafts for distribution to the tribal farmers. We also prepared a nursery of cashewnut seedlings and distributed them.

A training schedule in social forestry was organized for the tribals. They were supplied with plants of subabhool, and each tribal family planted subabhool on their lands. They could thus become self reliant in terms of fodder and fuelwood. Training was also imparted in horticulture. This has helped the tribals to develop fruit orchards of chikooes and mangoes. Besides every tribal family had established 'Nutrition Gardens' in their backyard.

I received valuable cooperation from Shri Harihar Save, Shri Vilas Save and Shri Shivaji Bhimara in development of the wasteland and in organising training and extension programmes. It was observed by us that if the wastelands were properly developed they could be utilized for growing fodder, fuelwood trees, timber trees and fruit trees. Non-government organizations could play a crucial role in imparting training to the villagers in developing wastelands.

Non-Browseable Crops and Plants

The tribals tie up their animals during the rainy season because at that time there is a rice crop in the field. Once the rice crop is harvested they let their animals loose for eight months. These animals return to their homes in the evening after grazing for the whole day. The tribals tie these animals in front of their huts. Formerly there was enough land for the grazing of animals and so the tribals could grow beans or grams after harvest of rice. These two crops were grown on residual moisture. Proteinous food was available due to beans and grams. Similarly, the plant residues from beans could be used as fodder. Both these crops, being legumes, enrich the soil through nitrogen fixation. But, nowadays the tribals have to keep their lands fallow due to the menace of stray animals.

To overcome the problem of stray animals we motivated the tribals to erect fences of cacti. The tribals who constructed irrigation wells had erected such fences and were able to grow vegetables and fruit trees. Some of the tribals who had bigger land holdings also erected the cactus fencing which enabled them to grow beans and grams. We arranged demonstrations to show that beans and grams could be grown successfully on the residual moisture. We also stressed the importance of putting a cactus fence. We undertook research to find out crops which are non-browseable. In Vada Taluka of Thane district one variety of Sesame *(Sesamum indicum)* namely *Kashal* was being grown successfully.

This variety was being grown in the rabi season. I went there specially to study this crop. The farmers told me that this crop was not browsed by cattle. We secured seeds of this variety and sowed it on the lands of tribal farmers around Kosbad Hill. We observed that this crop was not being browsed by the cattle. We arranged one demonstration of this crop and the tribals were convinced that sesame could be grown in areas where there was a menance of stray animals. We secured more seeds and distributed to a large number of tribal farmers. The tribals in the area are now growing sesame as the second crop after rice.

The second crop which could be grown in such areas is Safflower *(Carthamus tinctorius)*. The leaves of this crop have thorns and so the cattle are unable to browse them. However, in the beginning when the leaves are tender, the crop has to be protected for about two weeks. Thereafter when the thorns appear on the leaves there is no possibility of browsing. Safflower is a high value oilseed crop which could be grown on residual soil moisture.

Bottlegourd *(Lagenaria siceraria)* is yet another crop which is non-browseable. We demonstrated to the farmers the techniques of growing bottlegourd. The bottlegourd is a hardy vegetable plant which could be trailed on the ground. It requires hand watering.

An important fruit tree which is non-browseable is custard apple *(Anona squamosa)*. The seeds of custard apple are sown in polythene bags and planted at the onset of rains. This tree is very drought tolerant and does not require any irrigation. We introduced custard apple on the lands of tribals. The custard apple bears fruit in September and October and the tribals get cash income through their sale.

Karanj *(Pongamia pinnata)* is another plant which is non-browseable. We get non-edible oil from this plant which has industrial uses. The oil cake is used as a manure. The leaves of the *Karanj* are used as green manure. The branches are used as firewood. Besides, the flowers of *Karanj* contain considerable amount of honey and it promotes bee-keeping in the area. The tribals get good cash income through collection of *Karanj* seeds, which have a demand.

Vilayati babool *(Prosopis julifelora)* is a fast growing tree which is non-browseable. This tree is exotic and has been naturalised in India. The trunk and branches of this tree are used as firewood. The pods of this tree are sweet and liked by the animals. *Prosopis julifera* is a very hardy plant and could be grown in rainfed and dry areas. Besides, it could also be grown in saline and alkaline lands.

Australian babool *(Acacia auriculiformis)* is also a non-browseable plant. This is mainly used for firewood. The tree could be propagated by preparing seedlings in polythene bags and planting them at the onset of rains. This plant is very hardy and could be grown in the rainfed and dry areas as well as on waste lands.

Jetropha *(Jatropha curcas)* which is locally known as *Ratanjot* is also a non-browseable plant. The plant bears fruits which are of green colour. The seeds contain non-edible oil which is being used for soap making and for other industrial uses. This plant is also drought tolerant and could be grown in the desert areas. We tried to introduce the above crops and plants in the tribal areas where there was menance of stray animals.

9

FODDER CROPS

Paragrass

The tribal farmers begin their agricultural operations like ploughing, harrowing etc. when the rain starts in the first or second week of June. Thereafter, the rice is transplanted after puddling of the field. This work continues for a fortnight. In this way the bullocks continuously work for about one month and they are in need of good fodder. However, the local grasses do not grow during this period. The local grasses grow only in the month of August, so there is scarcity of green fodder during June and July. The bullocks are mostly fed on dried grass or rice straw.

There was need for a grass which would grow immediately after the start of the rains, so that green fodder could be available to the working animals. We introduced some grasses and started experimenting on them but we did not succeed.

Then, I had occasion to visit Aarey Milk Colony near Bombay. In this colony, cattle sheds were built to accommodate milking buffaloes. The milk from this colony was supplied to the city of Bombay. Here *Paragrass (Brachiaraia mutica)* was cultivated on the waste water from the cattle sheds. The speciality of paragrass is that it grows in the low lying areas having water logged conditions. This grass is exotic and was introduced in India from Africa, during the mid-fortys. Paragrass was known to be a water loving which could tolerate excessive water.

While going around the Aarey Milk Colony I went to one hill top and saw some stumps of paragrass growing very luxuriantly. The rains had started only a few days back and the paragrass had responded to the first showers. No other grasses were growing in the area. The length of the paragrass was about four feet. The grass was creeping on the ground and every node of the grass had roots. These roots penetrated into the ground and helped to supply nutrients from the soil to the paragrass. That was one of the reasons why the grass was growing so fast. There was no irrigation on this hill top which showed that this grass could survive the summers also. I observed, that the stumps sprout immediately after the first showers. I collected the stumps and brought them to the Kosbad Institute and planted them on the hill slopes where there was no irrigation. Since it was the rainy season there was no difficulty in planting as the soil was soft and had enough moisture. After a few days the stumps started sprouting. During that rainy season we obtained two cuttings.

During summer the grass stumps were in a dormant condition. However, immediately after the commencement of rains they started sprouting very fast and we were able to harvest them every fortnight. During the rainy season of four months we had about eight to ten cuttings. During that rainy season the working bullocks got green fodder for the first time. We distributed the cuttings of this grass to the nearby tribal farmers and they planted them on the uplands and on the bunds of the rice fields. Paragrass greatly helped the tribal farmers to solve the problem of green fodder during agricultural operations in June and July. In this way paragrass proved to be a boon to the farmers from rainfed and dry areas.

We had also observed that paragrass was an excellent soil binder. It helped to reduce soil erosion on the hill slopes. It also reduced erosion on the banks of streams and rivers. We also planted this grass on the banks of streams and rivulets, thereby providing excellent vegetative cover to prevent erosion.

Kubabhool to Subabhool

The Government of Maharashtra operated various schemes to achieve self sufficiency in milk production. The producers were given price assurance. Collection and processing of milk from rural areas were organized. Arrangements for transport, storage and sale of milk were also made. A programme of cross breeding of cows was undertaken.

Nutritious fodder plays an important rôle in milk production. With a view to evolve a programme for production of nutritious fodders in Maharashtra, a State level seminar was organized on the Government Dry Stock Farm at Palghar in 1965. The seminar was attended by milk producers, fodder experts and officers from the Departments of Animal Husbandry and Dairy Development. Shri P.K. Sawant, Maharashtra's Minister for Agriculture and Dairy Development presided over the seminar. Shri Manibhai Desai, Director of Bharatiya Agro-Industries Foundation (BAIF) was specially invited. The fodder experts presented papers about cultivation of leguminous fodders like lucerne, barseem, and cowpeas. Others presented papers on grasses like hybrid napier, and elephant grass. All these crops require irrigation.

I presented a paper on a fodder tree called 'Kubabhool' *(Leucaena glauca).* I stated that since ninety per cent of land in Maharashtra is unirrigated, we shall have to propagate thrust on fodders which could grow on dry lands. Kubabhool is one such fodder. The tree is highly drought tolerant and also grows on poor soils. The leaves contain 25 to 30 per cent crude proteins. Since the tree is leguminous it fixes atmospheric nitrogen and enriches the soil.

I narrated my observation about Kubabhool which we grew on the farm of the Kosbad Institute. The tree bears pods which have a large number of seeds. If the seeds are kept in hot water for about 15 minutes their germination improves. We used to prepare seedlings in polythene bags and distribute them among the tribal farmers. All tribal farmers use leaves of Kubabhool for feeding their animals.

Kubabhool is a native of Mexico. The English and Portugese brought this plant to India in the early part of this century. Due to its enormous seed bearing habit it has spread rapidly all over the country. However, no efforts had been made uptil now to undertake systematic plantations of Kubabhool. I stressed the need for undertaking such plantations in rainfed and dry areas in Maharashtra. I also suggested that, to begin with, such plantations could be undertaken on farms of animal husbandry and dairy development departments as well on lands belonging to voluntary institutions. Such plantations would prove to be demonstrations.

All the participants appreciated my paper. Shri Manibhai Desai told that Kubabhool should not be restricted to Maharashtra but should be propagated all over India. He collected large quantities of Kubabhool seeds from Kosbad Hill and propagated it all over the country. BAIF (Bharatiya Agro-Industries Foundation) and Sadguru Seva Sangh undertook projects of large plantations of Kubabhool at Chitrakoot in Madhya Pradesh, Manikgarh in Uttar Pradesh and Kutch in Gujarat. Two past students of the Kosbad Institute, with experience of growing Kubabhool, were deputed to these projects. I visited the Chitrakoot and Manikgarh projects and found that the Kubabhool trees had established well under rainfed and dry conditions of Uttar Pradesh and Madhya Pradesh. I was also told they performed well in the deserts of Kutch. The farmers from surrounding areas of all these three projects introduced Kubabhool on their lands.

Dr. M.S. Swaminathan, Director General, Indian Council of Agricultural Research appreciated the Kubabhool plantation at Kosbad Hill. He advised me to propagate this tree all over the country. He mentioned importance of Kubabhool in his convocation address at Bombay University. He advised the Professors and students to visit the Kosbad Institute to study this plant. Dr. Swaminathan also sent Scientists from Indian Council of Agricultural Research (ICAR) and Agricultural Universities to study plantation of Kubabhool. On their return all the scientists carried its seeds and introduced Kubabhool on lands of their Institutions. We established a seed farm of Kubabhool at Kosbad Hill and supplied seeds all over India.

I studied the research on this plant when I visited Australia in 1971. The Scientists at Cunningham Laboratory of Commonwealth Scientific and Industrial Research Organisation, Brisbane told me that they have collection of more than 300 varieties of *Leacaena glauca.* They have developed varieties suitable for dry areas in Australia. In Australia they sow the seeds of Leucaena in pastures, and when the plants are well established they allow the animals to graze in the pasture. They follow rotational grazing method which facilitates new growth of Kubabhool.

When I visited Philippines in 1980, I found that the whole country was covered with Kubabhool plantations. They call it 'Ipil - Ipil'. It was planted on both the sides of roads, waste lands and hilly lands. Ipil-Ipil was used for beautification of Metro Manila. The landless people get employment through harvesting leaves of Ipil - Ipil. The leaves are dried and used as feed for poultry, pigs and cattle. In Phillippines the tree is looked upon as a source for employment generation.

Dr. William Brubaker a plant scientist from University of Hawaii made a significant contribution in evolving fast growing and high yielding varieties of Ipil-Ipil *(Leucaena lucecephela).* One such variety was K-8 which became popular in South East Asian Countries. The Government of India acquired the seeds of this variety for introduction in India. Dr. M.S. Swaminathan sent me some seeds of this variety, for experimental plantation on our farm, we found that this variety grows fast and yields more leafy matter. We undertook seed production for distribution among farmers.

Shri Manibhai Desai undertook large plantations of K-8 variety on the farms of BAIF at Urlikanchan in Pune District. I attended a seminar on Kubabhool in 1981. Shri Manibhai Desai narrated the history of Kubabhool and mentioned the contribution of the Kosbad Institute in introducing this fodder plant since mid sixties. He further stated that since K-8 is a fast growing variety it would revolutionise Kubabhool plantation in the country. Kubabhool is a multi-purpose tree and could be used as fodder, fuel and fertilizer. The wood of Kubabhool could also be used as timber. A small hut was prepared out of Kubabhool wood which demonstrated that the tree would be useful in solving the problem of housing of the poor people. Available literature says that the leaves of this tree contains an alkaloid called *mimosine*. Due to this some people are reluctant to feed these leaves to their animals. However, he clarified that, from the experiment conducted at Urlikanchan they did not find any harmful effects on milch animals. He, however, cautioned that the leaves should not be fed in large quantities at a time.

Some participants proposed that the name 'Kubabhool' should be replaced by 'Subabhool' since the prefix 'Su' means 'auspicious'. This suggestion was accepted and it was proposed to name the plant as 'Subabhool'.

Shri Nanabhau Embadwar, Maharashtra's Minister for Forests who presided over the Seminar told me that it would be better if the name 'Subabhool' is announced by Shrimati Indira Gandhi, Prime Minister of India. He desired that we should carry one seedling to Delhi and request her to plant it and name it as 'Subabhool'. Accordingly I arranged for some seedlings from the nursery of the Kosbad Institute and Shri Nanabhau Embadwar carried them to Delhi. Shrimati Indira Gandhi planted one seedling in the garden of her residence on Safdarjung Road and named it as 'Subabhool'. From that time the name of Subabhool has become popular all over India. Shri Embadwar gave her some seeds of Subabhool from our Institute. She carried them to Rajasthan. She mentioned importance of Subabhool in her meetings and requested farmers to plant this fodder tree in the desert of Rajasthan.

I met Dr. Brubaker in an International Seminar on Agro-Forestry at Delhi in 1987. He told me that he had studied many plants of the world and had found that Subabhool *(Leucaena lucesophala)* was the most appropriate dryland tree due to its multipurpose uses like fodder, fuel, fertiliser and timber. He therefore concentrated his research on this plant and developed high yielding varieties, as his mission in life.

I told him the story of Kubabhool since 1965, and also its rechristening as Subabhool. He was very happy to learn that we were making efforts to introduce this plant for the last 20 years. He gave me his blessings and encouraged me to strengthen the efforts in introduction of Subabhool among the farmers of India.

Nutritive Fodders

I was invited by Community Aid Abroad, a donor agency in Australia to visit them in 1971. I got an opportunity to study fodder research in that country. Australia may be described as a country of pastures. The farmers here have developed the pastures so well that cattle and sheep graze all the year round in them. In the beginning I visited pastures around Perth, Adelaide and Melbourne. Fodder crops like clovers, and, lupins were grown here. I found that there existed a good mixture of grasses and legumes which helped to provide proteins to the animals. This improved the productivity of animals in milk, meat and wool. All these pastures were located in temperate regions.

When I visited Brisbane I could study about the research being conducted on tropical fodders. I met Dr. Humphray who was researching on fodders in the University of Queensland. He had written several books on fodders. He took me round the research farm and showed various species of fodders. I then visited Cunningham Laboratory of Commonwealth Scientific and Industrial Organization at St. Lucia. Here, also research was being conducted on tropical fodders. I met Dr. E.M. Hutton who was the director of the laboratory. Dr. Hutton had conducted extensive research on tropical fodders. He had made a significant contribution by evolving a new variety of Siratro *(Macroptelium atroperpurium).* I also met Dr. Date and Dr. Norris who were doing research on *Rhizobia*. They found out new strains of *Rhizobia* which were more efficient in fixing nitrogen.

I visited the experimental farm of the Cunningham Laboratory where existed a large collection of *Stylosanthes* varieties. Dr. Hutton told me that scientists from Australia had collected these varieties from around the world. They first experimented with a variety, viz., *Stylosanthes humilis* at their research station at Townsville which was located in the northern part of Queensland. This variety was found suitable for introduction in the pastures of Queensland. The farmers were impressed by its performance and introduced it on a large scale. It was popularly known as 'Townsville lucerne'.

Dr. Hutton told me that there was a need to introduce leguminous fodders like *Stylosanthes humilis* in grass lands of India. This would help to enrich the fodders thereby increasing the productivity of animals. However, he cautioned me that it was difficult to establish legumes under tropical conditions due to competition from local grasses. For that purpose continuous research would be necessary. He narrated his experiences in Australia where scientists had to conduct research for years to introduce leguminous fodders in their pastures. Dr. Hutton told me to take up this research on my return to India. He gave me about forty samples of seeds of various leguminous fodders. Dr. Date and Dr. Norris supplied me with *Rhizobia* cultures suitable for these fodders.

On my return to Kosbad Institute I laid out an experiment to test these fodders under agro-climatic conditions of Konkan region. The three varieties of *Stylosanthes* viz., *Stylosanthes humilis, Stylosanthes hamata* and *Stylosanthes gracilis* established well under heavy rainfall conditions. We also found that these varieties could grow in soils on hill slopes which are of low fertility. Siratro a creeper type of fodder was also found to grow well. There were some plants of Agati *(Sesbania grandiflora)* near the experimental plot of siratro. The creepers took the support of these plants and grew upto heights of 10 feet. Shri K. V. Raut, researcher at Kasbad Institute took keen interest in this experiment.

I sent my observations regularly to Dr. Hutton regarding the performance of these fodders. Dr. Date once visited our experimental plots when he came for a seminar in India. He found that the *Stylosanthes* varieties had good nodulation, which was a strong point for their adaptability under Indian conditions.

The most important observation about these fodders was their adaptibility to rainfed conditions and soils of low fertility. Due to these aspects there existed considerable scope for their introduction in rainfed and dry areas. We found siratro due to its trailing habit, if given support of trees, could produce high amounts ofbiomass. It was, therefore, suitable for agro-forestry programmes. We collected the seeds of these fodders for further experimentation.

During 1973, we observed that the old stumps of *Stylosanthes hamata, Stylosanthes gracilis* and siratro sprouted with the first shower of rains in June. This revealed that these fodders had perennial habit. The stumps of *Stylosanthes humilis* did not sprout but its seeds on falling on the ground germinated in such a large numbers that it gave the look of a green mat.

We established a fodder research section at Kosbad Institute in 1973 and invited Dr. M.S. Swaminathan, the then Director General, Indian Council of Agricultural Research for its inauguration. He was greatly impressed to see the experiments which we were conducting for the introduction of leguminous fodders in our grass lands. He told me that he had also recently visited Australia and brought one variety of *Stylosanthes scabra*. Dr. Swaminathan sent the seeds of this variety to us. We found that this variety grows erect and so it could withstand competition from grasses. It was perennial in nature and so once planted could stay for a number of years. Besides, the seeds of this variety did not shed from the plant which facilitated collection of seeds.

After returning to Delhi, Dr. Swaminathan sent a circular to the Vice-Chancellors of Agricultural Universities and Directors of Central Institutes of Indian Council of Agricultural Research pertaining to research work on leguminous fodders done at Kosbad Institute. On the receipt of this letter some Vice-Chancellors and Directors of Central Institutes visited the Kosbad Institute and witnessed the research work. We supplied them seeds and cuttings for introduction in their respective institutions.

The Community Aid Abroad was very happy to know about the performance of leguminous fodders. We requested them to send seeds of *Stylosanthes hamata* from Australia for their multiplication. Accordingly, they sent about 20 kilograms of seeds by air. We undertook seed multiplication programme at Kosbad Institute and produced about 2000 kilograms of seeds in that year. We informed Dr. Swaminathan about the availability of seeds of *Stylosanthes hamata.*

Dr. Swaminathan instructed the State Forest Departments to secure seeds of *Stylosanthes hamata* from Kosbad Institute for their silvi pastoral projects. In addition, Agriculture and Animal Husbandry Departments of various States also procured seeds from our Institute. Non-Government Organizations like Vishwaneedam at Bangalore, Karnataka, People's Multipurpose Society in South Arcot District of Tamil Nadu, Sadguru Sewa Sangh in Gujarat, secured seeds from us to undertake programmes of fodder development. Action for Food Production (AFRO), a non-government organization from New Delhi undertook projects for introduction of *Stylosanthes hamata* in West Bengal. I visited these projects in Midnapur and Sundarban areas. The development support team from Pune took up the project of fodder development in Rani Bannur area of Dharwad District in Karnataka which was a drought prone area. I visited the project and found that *Stylosanthes hamata* grew well under drought conditions also. Due to these visits I could study the performance of *Stylosanthes hamata* under various agro-climatic conditions in India. These experiences were useful to me while working on the Committee on Fodders of the National Wasteland Development Board in 1985.

I was very fortunate to get an opportunity to be exposed to research on leguminous fodders in Australia. I have observed that these fodders have played an important role in improving productivity of animals in that country. On my return to India I have tested these fodders under various agro-climatic conditions and found that there is a great scope for their introduction both in grasslands as well as wastelands of India.

10

WATER CONSERVATION AND MINOR IRRIGATION

Use of Stream for Irrigation

On the eastern side of the Kosbad Institute flowed a small stream. This stream originated from Kainad Hills and joined the Arabian Sea near Chikhale village and was about five miles in length. This stream flowed from June to October. After the rains, in October, the stream would dry up.

There was rice cultivation on both sides of the stream. However, no farmer was using water from the stream for irrigating rice. There was only one exception and that was of Shri Navsha Hadal who was a tribal farmer. Whenever there was gap in the rains, Shri Hadal used to construct a small earthern dam across the stream and divert water for his rice crop. He had prepared a permanent channel of about 500 feet length for taking water to his rice fields. The water flowing from this channel used to percolate and benefit the rice crops of other farmers. But these farmers did not make any attempt of putting an earthern dam across the stream. The crabs used to bore holes in this bund caused leakage of water. Shri Navsha used to keep day and night watch for control of crabs.

It was in August that I was taking the students of the Kosbad Institute to survey the rice crop in the nearby villages. Due to a long gap in the rains, the rice crop was withering for want of moisture. However, we found that Shri Navsha Hadal's rice crop was in good condition. It was due to the protective irrigation which he provided by constructing a small earthern dam across the stream, and laying a channel to take water from the stream. We enquired how had he prepared this channel on the contour even though he did not have any tools. He told us that the tribals have inherited the skills of carrying water from long distances. They take water to their rice fields from the streams flowing from the hills. They make water flow from one field to another. This provides fresh water to their rice crops and their crops are always in healthy condition. Shri Navsha suggested to us that we should put a small concrete dam across the stream so that there will be a permanent storage of water during the rainy season. This water could be diverted to the nearby rice fields of the tribal farmers. It would also be useful to the rice crop of the Kosbad Institute. Accordingly, we constructed a cheap concrete dam across the stream. This dam was found to be extremely useful for providing protective irrigation to the rice crop.

Construction of a small concrete dam had saved the efforts required for putting up earthern dams every year. The earthern dam also requires fresh earth every year and considerable top soil erosion takes place due to removal of earth.

We arranged one demonstration to show the utility of small concrete dams across the streams. A large number of tribal farmers attended this demonstration. They suggested to us that there were suitable sites for construction of more such dams on the same stream. Accordingly, we helped them to construct small concrete dams across the stream. Construction of small concrete dams across the streams was found to be a useful method for conserving rain water, and for assuring the rice crops of small and marginal farmers.

Irrigation Wells

This is the story of 1957. Shri Yashwant Jaeel, Deputy Secretary, Social Welfare Department, Government of Bombay visited the Kosbad Institute to see the tribal welfare work. We took him around the tribal hamlets and showed him the drinking water wells. These wells were constructed with the help of *shramadan* (voluntary help) of the tribals, students and staff members of our Institute. The Government gave financial support for the masonry construction of these wells. Shri Jaeel was very happy to see how these wells have helped to solve the drinking water problem of the tribals in the area. He told us that the Government of Bombay had a scheme to construct irrigation wells for the tribals. There was a subsidy of Rs. 1500 per well. Shri Jaeel further mentioned that very few tribals were taking advantage of this scheme. He requested us to motivate the tribals to undertake construction of such wells. We agreed to his proposal for construction of ten wells to start with. He promised us that he would send an amount of Rs. 15,000 for ten wells when he returned to Bombay.

We selected ten tribal farmers from the surrounding hamlets. We explained to them about the scheme of irrigation wells. They told us that they were unable to incur any expenditure for the construction of these wells as they did not have any cash income. We assured them that we would complete these wells within a ceiling of Rs. 1500, as students and staff of the Institute would offer voluntary labour in digging these wells. The tribals responded and agreed to join in the voluntary labour along with their families. We divided our students into ten batches and assigned one well to each. We purchased pickaxes, crow bars, spades and iron baskets. Every morning, students and staff used to go for digging these wells. The tribals with their families joined in the digging operation. Fortunately, eight wells struck water at a depth of fifteen feet. Two wells struck rock and had to be blasted by explosives. These two wells got water at a depth of about twenty feet.

The masonry work of these wells was done by local masons. We helped the tribals to instal low cost persian wheels for lifting water. These persian wheels were manufactured by the local smiths. The tribals purchased buffaloes to run these persian wheels. As there was no expense involved in digging the well, the cost of masonry construction, setting up of persian wheel, and purchase of male buffalo was possible within the ceiling of Rs. 1500. All the wells were ready by summer.

The rains in that year started in the second week of June. The ten farmers had sown their rice seeds on monsoon rains. However, after a week there was a gap in rains and the rice seedlings started drying up. Immediately the farmers irrigated their rice nursery with their persian wheels. Their seedlings could be protected from the drought conditions. These farmers also applied water at the time of puddling of the fields. This facilitated transplanting of rice in time. Whenever, there was a gap in the rains, they irrigated their rice crops. In this way, due to these wells, their rice crop became assured.

In the month of October all these farmers had prepared vegetable nurseries on the uplands. They consisted of egg plants (brinjal), tomatoes and chillies. We helped them to secure good seeds from the Department of Agriculture. We gave them training for preparing the vegetable nurseries. By the end of October, when the harvest of rice took place, they ploughed their lands, applied compost and cattle dung manure. They prepared ridges, furrows and planted their vegetables seedlings. We gave them guidance in cultivation of vegetables. In the month of January the vegetables were ready for harvest. There were weekly markets in the nearby townships. The women of these ten families used to get up early in the morning to carry their vegetables in bamboo baskets for sale in the markets. They bought a balance with weights and measures for the first time in their life. They acquired necessary skills for sale of vegetables. Uptill now the tribal women were going to the towns to purchase vegetables but, now they were going to sell their vegetables. They were very proud of this achievement.

We motivated them to save some cash which they were getting through the sale of vegetables. They opened Postal Savings Accounts. We told them that this saving would be useful to deepen their wells so that there could be more storage of water. This would help to bring more area under vegetable cultivation. In this way every farmer made a saving and deepened his well. Consequently, there was more storage of water and the production of vegetables significantly increased.

Introduction of vegetables generated cash income. It improved their purchasing power. Inclusion of vegetables caused significant improvement in their diet. The families got year round employment on their lands. Small irrigation wells are important instruments for generation of gainful employment among weaker sections.

Irrigation Wells and Pumping Sets

One day Shri K.V. Joshi, block development officer of Panchayat Samiti, Dahanu came to Kosbad Hill in October 1965. He was a follower of a sect called 'Warakari Sampradaya', who were devotees of God Vithoba of Pandharpur. They led a simple life and encouraged spiritual thinking. Shri Joshi was very committed to the service of tribals.

He told us that there was a scheme of construction of irrigation wells on the lands of tribals. However, he did not get response from the tribals. For that, he required cooperation from NGO's like us. I assured him that we would extend all cooperation in implementation of this scheme. I told him that we would go to the tribal farmers and explain the scheme to them. Accordingly, in November we went around Chimbave village and met the tribals. That was the threshing season of rice crop and so we could meet them on threshing yards. Ten tribal farmers showed their willingness to construct wells, since they had seen such wells on the lands of tribals in the Kosbad area.

We asked them what would be the suitable time for construction of wells. They told us that the work should be started during the current month itself. Further, they told us that during November, December and January they had stocks of foodgrains in the house, hence, the entire family could take part in digging these wells. Besides, in winter they could put in more hard labour. We agreed to their suggestion and arranged to supply necessary tools for digging the wells. The tribals have a certain inexplicable wisdom about sources of ground water. Accordingly they selected the spots on their lands. They performed *pooja* (worship) by applying *kumkum* (vermillion) to mother earth and breaking a coconut.

The students and staff of the Kosbad Institute also joined in the digging operations. All the members of tribal family including children joined the work of digging the wells. The engineers and overseers of the Dahanu Panchayat Samiti were regularly visiting the sites to take measurements of earth work. The *gramsevaks* (village level workers) were making payments of labour charges according to the norms of the scheme. Accordingly the digging portion of the wells was completed within three months. Water was tapped at a depth of 5 to 7 meters.

We helped the tribals in masonry construction. They did not have any experience in the past of such construction. We took them to the sites where bricks were prepared. They selected the bricks after checking their quality. They purchased cement from Dahanu Road. They brought the sand in their bullock carts from a coastal village called Chikhale. The masons from this village constructed these wells. The construction work was completed in May before onset of rains. In July, all the wells were full of water. However the pumping sets could not be installed as there was no electric power in the vicinity. We moved the Maharashtra State Electricity Board to install a transformer and electric lines. They agreed to our proposal.

According to the scheme there was a ceiling on the cost of installation of pumps. There was no provision of pipes and other accessories. Besides these wells were situated in the interior areas and the cost of erection of poles and electrical wires could not be accommodated in the scheme. Even though there was water in the wells, the tribals could not utilize it for want of pumps.

One day, a press reporter visited the Kosbad Institute and saw these unutilized wells. He reported this under the caption 'National Waste'. This news appeared in a leading newspaper all over India, and was read, amongst others, by Shri Roshanlal Aggarwal in New Delhi. He wrote me a letter stating that he represented a donor agency called Community Aid Abroad from Australia and would like to visit the Kosbad Institute to see these wells. Accordingly he came to Kosbad Hill from Delhi. I took him to Chimbave village where these ten wells were constructed. All the wells were full of water. However, due to lack of pumps the water could not be utilized. I explained the difficulties about installation of pumps.

On the advice of Shri Aggarwal, we prepared a project report in which the Government grant for pumping set and extra expenditure required for installation and purchase of other accessories was incorporated. Shri Aggarwal forwarded the project report to the Community Aid Abroad (Australia) in Melbourne. The organization was very kind to give immediate approval to this project. Shri Aggarwal made arrangements to send a cheque for meeting the extra expenditure required for installation of these pumps. We moved the Panchayat Samiti, Dahanu to secure pumping sets. We also approached the Maharashtra State Electricity Board to install the transformer in Chimbave village. This was the first occassion of installation of ten electric pumping sets at one time in a tribal village of Thane district.

After installation of pumps the tribal families made preparations on their land for planting vegetables. Since this was the first year we supplied them with seedlings of vegetables from the community nursery of the Kosbad Institute. We also gave them training in cultivation of vegetables. Vegetables like radish, brinjals and tomatos were ready for harvest within two months. The tribal women were packing their vegetables in bamboo baskets and taking them for sale to the nearby villages. They were getting cash incomes of Rs. 20 to 25 every day through the sale. Due to availability of fresh vegetables, the tribals in the surrounding villages also started purchasing vegetables. They included vegetables in their diet. Formerly their diet consisted only of cereals. We found that establishment of vegetable gardens helped to overcome the problem of malnutrition among the tribals.

This area was suitable for growing a fruit tree called chikoo (Sapota). There were large orchards belonging to landlords and other affluent people. These orchards were developed through the hard labour of tribals. However, the tribals did not have any opportunity to develop their own orchards. We therefore, motivated these ten families to introduce chikoo plantation on their lands, and supplied them with some chikoo grafts from the Institute's nursery. We gave them training in grafting of chikooes. The tribals prepared their own grafts and planted them on their lands. The chikoo tree requires four to five years for bearing fruits. Till that time the tribals grew vegetables as intercrop. We helped the tribals to sell their chikoo fruits in the Bombay market. Since chikoo trees in this area are bearing fruits all the year round, the tribal farmers received regular cash income through sale of fruits.

These ten wells became demonstrations of multiple cropping and fruit trees. The tribal farmers from surrounding areas were greatly impressed and took up construction of such wells. More than one hundred such wells have been constructed in five tribal villages of Kosbad, Waki, Chimbave, Gholvad and Chikhale. We organized a cooperative for supply of inputs like seeds, manures and fertilizers. Weekly markets were organized in surrounding villages for sale of vegetables and fruits. The landless artisans like potters, basket makers, carpenters and smiths were coming to the markets to sell their articles. Small grocery shops, tea stalls etc. were started near these markets. Tribal youths started small workshops for repairs of pumps, tools and implements. Trucks and power tillers were used for transport of fruits and vegetables.

The tribals who constructed wells have established beautiful chikoo orchards. They are now known as *Wadiwala* (orchard owner). The people who toiled as labourers for generations got the opportunity to develop their own orchards. Their purchasing power improved through sale of fruits and vegetables.They are now leading a life with self respect. The multiplier effects of their enterprise had also enriched the area by bringing technology and prosperity within the reach of all.

Surface Well

On the hill slopes of Konkan region rice, ragi and hill millets are the principal crops grown. Even though, the region does receive high rainfall, there exist intermitent gaps in its distribution over time. During such times the crops start withering and yields are hence adversely affected. Protective irrigation becomes necessary to save crops during such gaps.

Some of the rice fields in Kosbad Institute are situated on hill slopes. Early varieties of rice are grown on this land. We constructed a well on this land in 1965 to harvest rain water. The diameter of the well at the top was 60 feet and at the bottom 15 feet. The depth of the well was also 15 feet. The objective of keeping the shape of the well small was to enable the small and marginal farmers to construct such wells. A conical shape was given to this well so that there was no need for masonry construction. Only on the upper portion stones were arranged on slopes to protect from soil erosion. The whole operation thus was labour intensive and no outside material was required.

With the first shower of rains the whole water was diverted to this well. The water carried dried leaves and other refuse with it and got accumulated at the bottom. When the well became dry in summer good organic manure could be obtained which was utilized for rice crop.

Whenever there was a heavy shower of rain the well was filled with water. The water percolated to the rice fields which were situated at the lower levels. This also helped to preserve moisture in the soil during long breaks in rains.

Shri Mohan Parikh from Bardoli told us that he had developed a hand operated pump which was ideally suited to this well. This pump was very simple and could be operated by two persons. It was useful for irrigating the rice crop during gaps in rains.

This well was inaugurated by Shri Gopalrao Khedkar, Minister for Rural Development, Government of Maharashtra, and was named 'Surface Well'. Later, this name became popular everywhere.

We planted seedlings of glyricidia around this well. This helped to obtain green manure for the rice crop. We also planted mango and drumstick trees which gave fruits and vegetables. The leaf canopy of these trees helped to reduce evaporation of water from the well.

Demonstrations of this 'Surface Well' was also arranged. Shri Vasantrao Naik, Chief Minister of Maharashtra attended one such demonstration, and was greatly impressed by this project. He told us that this pattern of water conservation would be most appropriate for hilly terrain. He declared that the Government of Maharashtra would accept this pattern in programmes of drought relief. Accordingly, hundreds of surface wells were constructed in the hilly areas of Thane district. We found that a small water conservation device such as the 'Surface Well' was very useful for providing life irrigation to the rice crop.

Fish Ponds

During my visit to Israel in 1978, I had an occasion to study the rearing of fish. In Israel there are large number of tube wells. However, their water was brackish. This water was mixed with sweet water and was used to irrigate the cotton crop. The remaining water was utilized for rearing of fish.

Israel had developed a new technology for fish rearing in shallow waters. Since, the nation was short of water they found out that fish could be reared even in a meter of water. They converted crop fields into fish ponds by utilizing brackish water from nearby tube wells. There were large number of fish nurseries where fingerlings were prepared. These fingerlings were then put into fish ponds. Every fish pond was provided with an overhead container filled with fish food. This fish food had all the nutrients required for fast growth of fish.

When I returned back to India, I initiated an experiment of rearing fish in the Kosbad Institute. We selected one rice field which was of half acre size. This field was near a dug well from where the water could be easily pumped. The field was then filled with water and a level of one meter was maintained.

Due to sticky soil, there was less percolation of water from pond. In month of July, the fingerlings of carps were put in this pond. We prepared the feed for the fish according to the formula given by Scientists of Konkan Agricultural University at Ratnagiri. We installed a container to store the fish feed in one corner of the pond. The feed was regularly fed to the fish.

During rainy season, whenever, there was gap in the rain the water was pumped for filling the pond. After the rain was over in October, the water was pumped reqularly to fill the pond. We found the water which had percolated from the pond was going back to the well. In this way, there was recycling of water. We harvested the fish in the month of February. The fish attained the weight of 400 to 500 grams. There was a great demand for the fish in the local market. We found that if small ponds were prepared in the fields near the dug wells the water could be recycled for rearing of fish.

We held a demonstration, which was attended by Shri Sharad Pawar, the then Chief Minister of Maharashtra. The members of the State Planning Board who accompanied him were greatly impressed to see the rearing of fish in shallow waters. Some farmers from Thane District adopted this pattern and started rearing fish.

11

VILLAGE INDUSTRIES

Bee Keeping

We started the bee-keeping section at the Kosbad Institute in 1949. Shri N.S. Pandit was looking after this section. He was from Karwar District and had considerable experience in bee-keeping. The agro-climatic conditions were favourable for bee-keeping as there were forests around the Institute. Besides, there were orchards of chikoo, coconut, lemon, banana, papaya etc. In addition, there were cultivations of flowers and vegetables. Barring two months of heavy rains, namely July and August, there was plenty of flora for other ten months of the year.

Shri Pandit brought bee boxes from Shri Bapu Saheb Shende who was doing bee-keeping work at Pune. Shri Bapu Saheb was one of the pioneers in establishing the bee-keeping industry in Maharashtra. The *Apis indica* species of bees have a habit of preparing seven bee-hives and so seven frames are kept in the bee-box. The queen bee continuously lays eggs in these hives. On the lower side of the hive the eggs are laid while on the upper side honey is stored. For that purpose a separate frame of hive is kept for storing honey. This frame is called 'Super'. The Super frame is taken out and put in a centrifugal machine for extraction of honey. By this method, honey is extracted without disturbing the hive.

Shri Pandit arranged a demonstration to show how the bees could be captured from natural conditions and kept in the bee-box. There was one banyan tree in a hamlet called Varaipada. This tree had a hollow trunk. We all stood in front of it and Shri Pandit showed us the movement of bees indicating that there were bees inside the hollow trunk. We therefore, cut some portion of the trunk to facilitate removal of hives. Shri Pandit inserted his hand to touch the hives. The worker bees immediately came out and started hovering around us. Shri Pandit carefully cut out the hives with a knife and separated them from the trunk. He then gently tied them in the frames of the bee-box. There was a queen bee in these hives. She was unable to fly and tried to hide herself in the hives. He then placed the queen gate. The queen gate was prepared in such a way that only worker bees could pass through it, but the queen bee could not. All the worker bees entered the box. We left the box at the site till it was dark. Till that time all the worker bees that had gone outside had returned to their hives in the bee-box. We then carried the box to the Institute late in the evening. Shri Pandit kept this box on a wooden pole which was about one meter in height. On the base of this pole a small circular cement tank was constructed in which water was put. This water prevented ants from entering the bee-box. In this way we prepared twenty bee-boxes by capturing bees from surrounding villages. We constructed a wooden frame over these boxes and planted a flowering creeper called Antigonon *(Antigonon leptopus)*. The flowers of this creeper are pink and contain considerable amount of nectar. We named this apiary *Madhukunj*. We started giving training in bee keeping to the students of the Kosbad Institute and the tribal farmers. Shri Vishwanath Patil, an instructor from the Institute took great interest in bee-keeping and he acquired skills in this subject. Snri Pandit left the Institute and went to Mount Abu. Thereafter, Shri Vishwanath Patil was looking after this section. Many tribal farmers in surrounding villages have started bee-keeping. One bee-box used to give about eight to ten kilograms. of honey every year fetching about good subsidiary occupation for small and marginal farmers as well as landless people.

The Maharashtra State Khadi and Village Industries Board had appointed Shri Chawre as bee-keeping assistant. He did good work in popularising bee-keeping among the farmers of Dahanu Taluka. We arranged one meeting of the bee-keepers and impressed upon them that there was a need to start a Bee-keeping Cooperative. Accordingly the 'Dahanu Taluka Bee Keepers Cooperative Society' was established in 1965. Shri Gajanan Pandurang Churi from Bordi was elected as Chairman of the Society. About 2000 bee-boxes have been established under the auspices of this Society. The Society celebrated its Silver Jubliee on 21st April, 1990.

In this way a small bee-keeping centre which we started in 1949 at the Kosbad Institute has promoted bee-keeping among small and marginal farmers and also landless people, in the region.

Neera and Palm-Gur (Jaggery) Industry

There were date-palm trees on the farm of Kosbad Institute. We were tapping these trees for extraction of *neera* (Palm-Juice). We were also preparing palm-gur (jaggery) from *neera*. This village industry was first started at Bordi in 1939 by Shri Gajanan Naik. Since then the tradition of this village industry has been continued in this area.

During pre-Independence days Shri Gajanan Naik was staying in Sevagram Ashram of Mahatma Gandhi. Once he asked Gandhiji "what work I should do?" Gandhiji replied "there are large number of date-palm trees round about Wardha. Extract *neera* from them and prepare palm gur." From that time Shri Gajanan Naik devoted his life for this village industry.

After Independence Dr. Rajendra Prasad, President of India specially appointed Shri Gajanan Naik as Palm-gur Adviser to Government of India. While working in this post Shri Naik used to take only one rupee as honorarium. He travelled by third class in railways. He used to come to Kosbad Hill on foot from the Gholvad railway station. Even though he was a freedom fighter he never accepted any pension or other benefits. The All India Palm-Gur Training Centre at Dahanu was founded by him under the auspices of Khadi and Village Industries Commission. He was a very committed person and believed that *neera* and palm-gur were his life's mission.

There were about 300 date palms on the farm of the Kosbad Institute. We used to tap 150 trees for extraction of *neera*, following the Bengali method of tapping. Shri Gajanan Naik brought skilled workers from Bengal to introduce this method in this area. These skilled workers trained the local tappers. The local method of tapping was defective as it damaged the crown resulting in mortality of trees. In the Bengali method, tapping was done in such a way that the crown was protected and there was no possibility for damage of the trees. The Bengali method therefore became very popular in the region.

We used to send the *neera* for sale to Gholvad, Bordi and Dahanu Road. Besides, we were giving *neera* regularly to the tribal children of our Ashram School. We found great improvement in the health of these children. There are a large number of date-palm and sago-palm trees in the States of Tamil Nadu, Andhra Pradesh, Orissa and Bengal. People from these States extract *neera* from these trees and prepare palm gur. It is an important village industry, supplementing the incomes of almost every family. Shri Gajanan Naik used to say that Maharashtra also has great scope for starting this village industry. Shri Naik demonstrated that sugar could also be prepared from *neera*. He set up such a machine in the All India Palm Gur Training Centre at Dahanu.

Carpets and brooms could also be prepared from the leaves of the date palm. The landless agricultural labourers were practising this village industry in this area. All the parts of the date palm tree could be used as fire wood. The date palm is a truly multi-purpose tree.

We undertook a research project for improving the productivity of date palm trees. We experimented with both organic manures as well as chemical fertilisers. We found that the productivity of the trees could be improved by applying low cost green manures like glyricidia *(Glyricidia sepium), karanj (Pongamia pinnata),* neem *(Azadirachta indica)* etc. We therefore, regularly applied green manures to all the date palm trees at the onset of rains. The Khadi and Village Industries Commission had given financial support for this research project. We undertook plantations of date palm trees on the hill slopes and waste lands. For this purpose a simple technique was applied. We first prepared the seedlings in the polythene bags and transplanted them at the onset of rains. About one thousand seedlings were planted in one hectare. Cactus fencing was prepared around the plantations to protect them from the stray animals. The seedlings were so hardy that they did not require any watering during the summer. For the first three years the growth of the seedlings was slow, afterwards they grew very fast. We applied a mixture of organic manure and chemical fertilizers to the seedlings which enabled them to attain good growth. The trees came to maturity within ten to twelve years.

From our experience we found that date palm is a very useful tree. We get *neera* and palm gur both of which are rich in nutrition. Besides, carpets, brooms and other articles are prepared from the leaves, providing employment to the landless people. The trunk and dried leaves are used as firewood by the poor people. The most important aspect of the date palm is its drought tolerance. It can be grown in the rainfed and dry areas. It can also grow on soils of low fertility. It is the most appropriate tree for waste land development.

Village Industries

The tribals in Thane district grow two oilseed crops viz., niger *(Guizotia abyssinica)* and Sesame *(Sesamum indicum)* on uplands. They sold the oil seeds to local traders at a cheap price as there were no crushing facilities. The tribals hardly consume any oil due to the lack of purchasing power. We therefore installed a bullock driven crusher in Kosbad Institute. The Khadi and Village Industries Commission gave financial assistance for this crusher. We informed the tribals that we have made arrangements for extraction of oil on this 'Community Crusher'. On hearing this the tribal women started coming in large number with their niger seeds. Only labour charges had to be paid for crushing their oil seeds. Many a times they did not have cash for payment. In such cases we purchased their oil cake worth the crushing charges. This cake was utilized for our dairy animals. Due to an assured facility for oilseed crushing, the area under oilseeds increased significantly. The tribals who hardly consumed any oil now included oil in their diet. We found that setting up of the community crusher had helped the tribals to become self reliant in oilseeds production.

The landless artisans in the area prepared baskets from bambooes. These baskets were required on a very large scale for packing chikooes (Sapotas). The bambooes were in short supply and so the artisans had to purchase them at very high prices from traders. We therefore, took up a programme of introduction of bambooes among tribal farmers. For that purpose we prepared a nursery for supply of seedlings. We arranged for the training of tribal farmers in cultivation of bambooes as this was a new crop to them. The tribals planted bamboo seedlings on their lands. The bambooes started yielding after five years and became a good source of income for the tribals. The artisans also got bambooes locally at reasonable prices. The village industry of basket making was greatly promoted through bamboo plantations.

There were date-palm trees in the surrounding villages. The tribal women prepared carpets and brooms from leaves of the date palm trees. It provided gainful employment to tribal women during lean months. Date palm being xerophytic could be grown in soils of low fertility under rainfed and dry conditions. We prepared a nursery and distributed seedlings among the tribals, which they planted on their wastelands. We realised that the success of village industries such as oil crushing, basket making, preparation of leaf carpets and brooms depended on local availability of raw material. The wastelands in the area could be utilised for the production of raw materials required for village industries.

The agro-climatic conditions of Thane district were congenial for growth of mulberry. We therefore decided to conduct an experiment on sericulture. The Maharashtra State Khadi and Village Industries Board was the implementing agency for sericulture in the State. The board had its training centre in sericulture at Panchagani in Satara district. We deputed one of our workers from Kosbad Institute for training. In the meanwhile we brought cuttings of Mysore-5 variety of mulberry from sericulture farm at Panchagani. This variety was evolved by Central Silk Board at Mysore and was high yielding. We planted mulberry on half an acre. The growth of this variety was good and within five months it started yielding leaves. A small hut was prepared for rearing of silk worms. Equipment like wooden trays, stands were prepared by local carpenters while the *chandrikas* (circular bamboo trays) were prepared by local artisans. We brought disease free layings (silk worm eggs) from Panchgani. Necessary steps, like timely feeding of mulberry leaves to worms, protection of worms from disease and oozy fly were taken. The cacoons were ready within one month. We were able to take 4 crops within a year and the yield of cocoons was 150 kilograms per acre. Being the first year, yields of cocoons was less. However, in subsequent years the cocoons produced higher yields. These cocoons were purchased by Maharashtra State Khadi and Village Industries Board and sent for reeling at their centre in Wai in Satara district.

A demonstration of sericulture was arranged for the farmers in Thane district. Sericulture was a new village industry in the district. Some farmers in Dahanu and Palghar Taluka expressed their desire to plant mulberry for rearing silkworms. We made arrangements for training of these farmers through the Maharashtra State Khadi and Village Industries Board. These farmers planted mulberry and started rearing silkworms. The yield of cocoons were very good. However, the farmers faced difficulty in getting disease free layings (DFL) on time. These layings had to be secured from Panchagani which was about 300 kilometers from their place. Regular supply of DFL is therefore necessary for rearing of silkworms. The farmers faced another difficulty concerning the selling of cocoons. The Maharashtra State Khadi and Village Industries Board used to purchase the cocoon and sent them for reeling at Wai which was at a distance of nearly 250 kms. But many a times due to the delay in procurement spoilage of cocoons occurred. It is therefore, necessary to establish reeling centres in the production areas itself. We found that success of sericulture in non-traditional areas depended on the creation of infrastructure like grainages and reeling centres.

The coastal areas of Thane district had plantations of coconuts. However, nobody was using fibre of coconuts. We therefore, decided to start a village industry of coir. For that purpose we brought a hand operated machine from Kerala for preparing ropes. The coconut husk was brought from nearby villages of Bordi, Gholvad and Narpad. The students of the Kosbad Institute were given training in preparation of strings. We also trained the students in preparing various other fibre products. We found that there was a great scope for utilization of coconut fibre in non traditional areas.

We started hand pounding of rice in Kosbad Institute and purchased hand operated haulers. Tribal women from the nearby areas brought their paddy for hauling. The hand pounded rice contains more bran as compared to milled rice and so it is more nutritious. If community hand operated haulers are organized it would help to get nutritious rice. This would greatly help to overcome the problem of malnutrition among the weaker sections of the society.

We organized a unit of *Amber Charkhas* (spinning wheels) at Kosbad Institute. The tribal women were given training in running the *Amber Charkhas.* After their training they started spinning on the *Charkhas.* They were paid the necessary wages. We attached a creche to this unit which facilitated the tribal women to keep their infants. The yarn prepared in the unit was sent to Kora Gramodyog Kendra, Borivali for weaving. *Amber Charkha* provided gainful employment to the tribal women during the lean months of the year.

We set up a small mechanical workshop at the Kosbad Institute for training students. The students acquired skills in repairs of electric motors, power tillers and other agricultural machinery. After this training they started their own workshops in their villages. We found that setting up of mechanical workshops in the villages helped to generate gainful employment for the youth.

12

INTERNATIONAL COOPERATION

Cooperation of International Organizations

The Government of Maharashtra entrusted us the responsibility of conducting training in the applied nutrition programme. This programme was undertaken with the support of UNICEF. The objective of this programme was to overcome the problem of malnutrition through production and consumption of protective foods. Vegetables and fruits consist considerable amount of vitamins and minerals and so the programme entailed growing of vegetables and fruits. UNICEF therefore supported us in establishing seed farms and nurseries for the supply of seeds and plants to the beneficiaries under this programme. Ten villages were selected in each applied nutrition block. We conducted training of functionaries such as *gramsevaks, gramsevikas, balsevikas* and extension workers. The members of Mahila Mandals (Women's organization) also attended the training, and were given practical training in growing vegetable and fruit trees. We also gave demonstrations for preparing low-cost recipes from vegetables and fruits. The theme of the training was 'learning by doing'.

We took the trainees to see the nutrition gardens established in the backyards of the surrounding tribals by us. We introduced drought tolerant vegetable trees such as moringa and *Sesbania grandiflora* and fruit trees like Mango, custard apple, *amla, ber* and jackfruit. We also informed them about growing papaya and bananas on wash water. The nutrition gardens created great confidence among the trainees and led to them establishing the same in their villages. We supplied them seeds and plants too from our seed farms at Kosbad Institute.

I prepared few slides on Applied Nutrition Programme, which the UNICEF Representative in Bombay, Shri Mulgaokar, sent to New York. UNICEF made copies of those slides and sent them all over the world. We tried to design applied nutrition programme on principle of self-help. The people from weaker sections produced and consumed the protective foods consisting of vegetables and fruits. This has greatly helped to overcome the problem of malnutrition especially among the vulnerable sections like pregnant women, nursery mothers and children. UNICEF Representatives from all over the world visited Kosbad Institute to see the programme which imparted considerable encouragement to us.

A voluntary organization from Australia 'Community-aid Abroad' has given valuable support to the Kosbad Institute in carrying out various agricultural projects for the welfare of tribals. This organization was founded by Father Tucker to help the poor people in developing countries. It was first started in Melbourne and later its branches spread all over Australia. I was invited by this organisation to visit Australia in 1971.

During my visit I met various groups from Universities and social organizations and explained to them about the problems of weaker sections in India. I also showed slides about the various projects like education, health and agriculture carried out by us amongst the tribals. They were impressed to see these action oriented projects.

The groups and organizations I met showed me various programmes conducted for fund raising in their cities. In one of the school in Melbourne city, students undertook a programme of "walk against want". They first assembled on a ground on Sunday. Every student carried a plaque in which an appeal was made to help the poor people. The citizens of Melbourne sponsored this programme. Every student received a donation according to distance he walked. About ten thousand Australian Dollars were collected through such donations. I met some students after this programme and asked them about their impressions. They told me that they read about plight of poor children in developing countries, many of whom could not even attend school as they were hungry and malnourished. The children therefore considered it their duty to help these unfortunate children. The ten thousand dollars raised through this programme would be sent to one school in Gujarat to support the feeding programme of tribal children. I complimented them on their efforts. In the city of Adelaide a women's organization arranged a programme of "Slim-In". About 200 lady participants declared reduction in their weight and the sponsors declared donations. The fund raised through this programme was to be sent for the welfare of women among weaker sections in India.

The Community-Aid-Abroad was running stores of khadi and handicrafts imported from India. The profits accrued from these stores were sent for running projects of education and rural development in India.

The Community Aid Abroad was very fortunate to have devoted

workers like Mr. David Scott, Mr. Neil O Sullivan and others. I came in close contact with Mr. Neil Sullivan and found in him a deep commitment to solve the problems of the poor in developing nations. He had studied in depth about the causes of drought in Africa. He had also delved deeply into the social problems of different communities in developing nations. We received very valuable guidance from him in our projects at Kosbad Institute.

Shri Roshanlal Agarwal, the representative of Community Aid Abroad in India was also conversant with the problems of poor. He advised me to give emphasis on construction of small irrigation wells on land of tribals as these wells would enable them to raise multiple crops and provide gainful employment on their lands. He also advised me to keep an objective of providing *dal-roti* (pulse curry and bread) to a tribal family. He however felt that this was also a difficult task to achieve. He was of the opinion that people who talked about development of districts or States could not achieve much in concrete terms. He therefore appreciated my approach of selecting a small area and doing comprehensive rural reconstruction work.

Mrs. H. Bedi, Representative of Community Aid Abroad in India introduced new social approaches in implementing the projects sponsored by the voluntary organizations. We were undertaking various agricultural development projects under the auspicies of the Agro Industries and Tribal Welfare Foundation at Bordi. We organized a training camp of the tribal women in social forestry. The technical component of our project such as identification of proper fuel wood plants, raising and care of the plants was quite strong. However, the participatory approach of the training was weak. We tried to show visuals like films-strips, slides etc. in order to evoke interest, but, the tribal women seldom responded to our questions. Mrs. Bedi used motivational techniques like folk songs, folk lures and folk dances which considerably helped in creating confidence among the tribal women and henceforth started taking active participation in our training programme. At the end of the programme they decided to form a women's organization to tackle the problem of fuelwood. Later the women also organized themselves to combat malnutrition by setting up Nutrition Gardens. We found that this motivational techniques were very important for effective participation in the rural reconstruction programmes.

Dr. John Staley, Representative of OXFAM in India visited the Kosbad Institute very often. He supported a programme for conservation of rain water. We devised a conical shaped well on the hill slopes for that purpose. The water from this well was utilized for protective irrigation of the rice crop during long gaps in monsoon rains. He also supported a programme on construction of small irrigation wells among the tribals. This helped the tribals to raise multiple crops like, vegetables, fruits, fodders, thereby getting gainful employment on their lands.

The students from the University of Notre Dame in United States were involved in raising funds. For this purpose they fasted for a day every month. The collection raised by this programme was sent to the Kosbad Institute. We utilized these funds for feeding the children from the *Balvadis*.

In 1976 under the Colombo Plan I was sent to the United Kingdom to study the British Agricultural Education System. I studied piggery at the Staffordshire College of Agriculture. Mr. John Mathews who was the Head of Piggery Department, was considered a great expert in the field. I requested the British Council to depute Mr. Mathews for training the staff at our Institute. Accordingly the British Council was kind enough to spare the services of Mr. Mathews for four weeks. He imparted training to the staff at the Institute to establish a good piggery.

Mr. Bala Saheb Kher, former Chief Minister of Bombay State was appointed High Commissioner of India in the United Kingdom. He was aware of the tribal development activities being conducted at Kosbad Institute. He requested Lady Cripps, wife of Sir Stafford Cripps, Chancellor of Exchequer, to specially visit the Institute and to see the tribal development work being conducted. We showed her the transplanting of rice being done by the tribal girl students. She also expressed her desire to do transplanting and proceeded to do the same by entering the muddy field and practising transplanting with the tribal girls. She was impressed to see that the tribal girl students at the Kosbad Institute were learning agriculture and animal husbandry through work experience projects.

Training of American Peace Corps

In the year 1966 I received a letter from Miss Helene Moose that I should come to Bombay to meet her. She was working in the office of American Peace Corps in Bombay. When I met her she told me that before joining the peace corps she worked for some years on rural health at the Mauni Vidyapeeth of Gargoti, in Kolhapur district of Maharashtra. She identified with Indian Culture and used to consider herself an Indian. Dr. Sita Poovahia a close friend of hers had suggested her to write to me. I knew Dr. Poovahia as Director of United States Educational Foundation in India at Bombay. This foundation had granted me the Fulbright Scholarship in 1962 to pursue higher studies in the United States. On my return to India from United States, Dr. Poovahia visited the Kosbad Institute and was greatly impressed to see the educational and agricultural development work which we were doing among the tribal people. She suggested to Miss Helene Moose that the Kosbad Institute would be an ideal site for imparting training to American Peace Corps.

One day Helene Moose visited Kosbad Hill. On seeing the rural reconstruction work which the Kosbad Institute was doing among tribals she selected it for imparting training to American Peace Corps. On return to Bombay she moved the Government of Maharashtra to approve the Institute for training of Peace Corps. The American Peace Corps were established during the tenure of President John F. Kennedy. The objective of this organization was that the volunteers of the Peace Corps should develop a spirit of social service by actually working with the people of developing nations.

The first batch of volunteers came for training in June 1967. The batch consisted of both men and women. They were to work in the area of health for a period of two years. The volunteers had to first learn the local language. Those who came to Kosbad had to work in the State of Maharashtra and were hence learning Marathi. They were given training in rural health. They were also given practical training in sanitation like preparation of soak pits, toilets, compost pits etc. Besides this, they were also trained in immunization and first aid. At the end of the training every volunteer presented his or her demonstrations in Marathi.

The Officers of the American Peace Corps in Washington visited the Kosbad Institute to observe the training process. They were greatly impressed by the atmosphere of the social work at the Institute. On their return to the United States they recommended that the training of the Peace Corps in Maharashtra be conducted at Kosbad Hill. Accordingly, we were given the responsibility to train the batch of volunteers who were to come in November 1967 for working in area of Applied Nutrition Programme. The first part of the training for two months was organized at the School for International Training at Brattleboro in the State of Vermont in the U.S.A. The American Peace Corps requested me to go to Brattelboro for teaching Marathi and imparting training in applied nutrition programme to the volunteers. Accordingly, I went there in September 1967. This batch of volunteers consisted of couples. The objective of this approach was that both husband and wife could work together in India. Mrs. Anne Janeway was the training organizer. Mr Ray Clarke was the Organiser for training in languages. Latest techniques were used for imparting training in linguistics. This helped the volunteers to quickly acquire knowledge of Marathi. Many of them even started speaking Marathi within two months

After completion of two months training in United States the volunteers came to Kosbad. Instead ofkeeping them in hostel we arranged their stay with the families of our staff members. This helped them to internalise Indian customs. The volunteers also became fluent in spoken Marathi as they had to speak in Marathi with the families of the staff members. We trained them in establishment of Nutrition Gardens in the backyard, preparation of recipes from vegetables and fruits, and simple low cost processing methods. The volunteers were taken to the tribal villages and each couple was assigned a tribal family. They gave demonstrations about various aspects of applied nutrition in Marathi. This created self confidence among them. After completion of training the volunteers went to work in the applied nutrition blocks in the State of Maharashtra. They worked there for two years. The Director of the Peace Corps in Bombay desired that I should visit these blocks and guide the volunteers in their work. Accordingly, I was visiting these volunteers regularly. I found a sense of commitment among all these volunteers. This was also appreciated by the villagers.

Mr. Guard Shular and his wife Mrs. Van Shular were working in the Applied Nutrition Block in Amravati. Both were more than sixty years of age. They came with the sincere desire of doing social service in India. During the training period, Guard dressed as Santa Claus on X'Mas day. He went around in a bullock cart and distributed gifts to the tribal children. This was their first look at how the festival of X'Mas is celebrated. Guard was working very hard to make the applied nutrition programme a great success. He was moving from house to house and convincing the people to plant vegetables and fruit trees in their backyard. The summer of Amravati is very hot and unfortunately Guard had a sun stroke. He was immediately removed to the Medical College Hospital at Nagpur. The American Embassy in Delhi specially sent a team of doctors by air to Nagpur. Despite their efforts, Guard could not be saved. Mrs. Van Shular went back to the United States after the death of her husband. Her sons who were working on high posts asked her to stay with them. Mrs. Shular told her sons that she would go back to India and complete her mission of two years, as a homage to the memory of their late father. Accordingly she returned back to India, served for one more year and completed her mission. This is a moving example of commitment of selfless service.

One more batch of American Peace Corps Volunteers came for training to the Kosbad Institute. They were to work for two years in the area of tribal agriculture. This batch also consisted of couples. Since the Kosbad Institute was situated in the tribal area we could organize training relevant to tribal agriculture. We trained them in growing high yielding varieties of rice, pulses, oilseeds, horticulture, fodders and agro-forestry. Mr. Joseph Blatchford, World Director of Peace Corps in Washington specially visited Kosbad Hill to see the training of the volunteers. The volunteers showed him the various agricultural skills that they had acquired during training and they also demonstrated these skills to the tribal farmers. Mr. Blatchford was impressed with the training programme at Kosbad Hill.

After completion of this training the volunteers went to the tribal areas and stayed there for two years. I used to visit them occasionaly. Once Dr. A.U. Sheikh, Secretary, Rural Development Department, Maharashtra came to Kosbad Hill and told me that he wanted to see the work of the Peace Corps volunteers. I suggested Talwada in Jawahar Taluka where Mr. Ron Fultan and his wife Cynthia were working. We went straight to Talwada without any intimation. When we reached their house we were told that they had gone to a nearby village. We requested a teacher from the Ashram School at Talwada to accompany us to that village. We saw that Ron was applying fertilisers to the rice field and Cynthia was transplanting rice. Mr. Sheikh was very much impressed to see the dedication of the Peace Corps volunteers in transfer of agricultural technology. They were demonstrating the cultivation of high yielding varieties by actually working in the field. The American Peace Corps Volunteers did very good work in enhancing tribal agriculture. In recognition of their services a felicitation function was held at the Kosbad Institute, at the end of their term. The American Ambassador in India, Mr. Kenneth Keating specially came from Delhi to felicitate the volunteers. He met every volunteer and discussed with them about their work during the last two years. Mr. Keating announced that the work of these volunteers in tribal areas was rated as the best in India. Swami Anand and Acharya Bhise also attended the felicitation function of these volunteers.

Mr. George Conk and his wife Margo had to leave India two months before the end of their term as they wanted to join the University for further studies. They came to see me before they left. They told me that they had saved Rs. 3000 and wanted it to be utilized for some good purpose. We bought one centrifugal pump and installed it on the well of one tribal farmer Shri Barkya Magi. The pump was inaugurated by Mr. Kenneth Keating. Due to this pump the tribal farmer could provide irrigation to his rice crop during break in rains. His rice crop became assured due to protective irrigation. He grew vegetables after the rice harvest and gained employment throughout the year on his land. It is a example of how through a small gift a tribal farmer obtained gainful employment.

I had an opportunity to observe the work of American Peace Corps

Volunteers in the State of Maharashtra. They were staying and working in very interior areas sacrificing the comforts which they had in their country. I found a commitment for work among these volunteers. Besides they developed dignity of labour. These volunteers tried their best to bring about friendship between the rural people of India and people of the United States.

Visit of Ambassador of West Germany

Many visitors from abroad used to come to Kosbad Institute to see the development work being conducted by us among the tribals. One day I received a letter from Mr. Gunther Dheil, Ambassador of West Germany, in which he expressed his desire to visit our Institute. I immediately replied to his letter stating that we would be happy to welcome him. Subsequently, Mr. Gunther Dheil, his wife Mrs. Helga Dheil and Consul General of West Germany in Bombay Mrs. Schottle visited the Kosbad Institute.

We took the visitors to tribal farmers and showed to them the project on small irrigation wells. The farmers as a result of this project started growing wheat crop for the first time. Formerly the farmers grew only rice crop in monsoons. After harvest of rice the land had to be kept fallow. The availability of irrigation enabled the farmers to cultivate the wheat after harvest of rice. With the introduction of wheat the farmers now included *chapati* in their diet thereby enriching it. Formerly they consumed only rice. Some of the tribals also showed the ambassador a variety of irrigated vegetables grown by them. The farmers stated that the cereals in their diet were supplemented by vegetables. The ambassador was impressed to see the improvement in the diet patterns of the farmers due to growing of wheat and vegetables. The ambassador was also taken to a farmer who had raised a small orchard of chikoo fruits.

One farmer informed that he got regular cash income through the sale of fruits to Bombay market. The ambassador was happy to see the improvement in purchasing power due to the introduction of horticulture.

The ambassador enquired about the cost of construction of one well and the installation of pump. I informed him that while the construction of a well costs about rupees five thousand, the pump could be installed in about two thousand rupees. I further told him that the tribals in this area had become owners of their land in 1951 after the passing of Land Reforms Act popularly known as 'land to the tiller'. Each family now owned about one to one and half hectares of land. The tribals grew only one crop of rice which was exposed to monsoon vagaries. After the harvest of rice the family did not have any gainful employment for eight months of the year. I further told him that in case a small well and pump was made available it 'provided protective irrigation' to their rice crop and hence saved the crop from vagaries of the monsoon. After the harvest of rice they could grow wheat and vegetables. Besides they could also raise a small orchard of fruit trees. The multiple cropping provided them gainful employment on the land and hence curbed migration towards urban areas. The purchasing power of the family also improved. I told him that provision of small irrigation wells was an important programme for the generation of gainful employment among the weaker sections like tribals.

On return to Delhi I received a letter from Mr. Gunther Dheil enclosing a cheque of Rs. 7000. He wrote in this letter that his wife Helga would be shòrtly celebrating her 60th birthday. She had decided not to incur any expenditure on entertainment but to utilize the amount as a donation to a poor tribal family for the construction of a well and installation of pump. We acknowledged the cheque with thanks.

A tribal farmer from Dalwipada hamlet was selected for whom a well was dug. The staff and students from our secondary school participated too in the digging operation. The tribal farmer and his family members too put in their hard labour. Water was tapped at a depth of about five meters. The masonry construction was done by bricks and cement and a centrifugal pump was installed on this well. Ambassador Dheil was informed on the completion of the project. We also wrote to him that we proposed to fix a small marble plaque to acknowledge the donation given by Mrs. Dheil. Ambassador Dheil replied that inauguration of the well was to be performed by the tribal farmer and his wife. He also wrote to us that 'Helga's Brunnen' (Helga's well) be written on the small marble plaque.

According to their wishes we inaugurated this well and fixed up the marble plaque. In a novel manner through a small birthday gift, Mrs. Dheil helped a tribal family to attain self reliance.

Study of Agriculture in Other Countries and its Application in India

The United States Educational Foundation in India (USEFI) granted me travel grant under the Fulbright Programme to pursue higher studies in U.S.A. in 1962. I obtained Master of Science degree in Extension Education at the Kansas State University in 1964. I stayed in United States for a period of about twenty months. During this period, I got an opportunity to study the American Agriculture. I visited some apple orchards in Ohio State. The farmers told me the history of introduction of apples in United States. Mr. John Chapman, who was born in 1775 in

Leominister, Massachusetts, devoted his life to introduce apples in Mid West. He did the work of creating apple orchards for over fifty years. The people of United States called him 'Johnny Appleseed' for his contribution to the introduction of apples in United States.

While at the Kansas State University, I also got an opportunity to read a book *World was my Garden*, a biography of Mr. David Fairchild. David was son of President of Kansas State University. For over sixty years, he did the work of exploring plants from all over the world and introducing them in United States. Prominent among them were soyabeans, wheats, dates, cotton, vegetables, fruit trees and flowering plants. He started a Division of Foreign Seed and Plant Introduction in United States Department of Agriculture. I was considerably inspired by the contributions of Johny Appleseed and David Fairchild in the field of plant introductions.

I studied cultivation of asperagus at Kansas State University. On my return to India I carried seeds of asperagus of Mary Washington variety from Kansas State University. I initiated an experiment of introducting asperagus at the Kosbad Institute. We observed that even though it was a temperate vegetable, it could be grown successfully under tropical conditions. Seeing our experiments some enterprising farmers in the area started growing asperagus on a commercial scale. I studied the cultivation of vegetables in glass houses in U.S.A. These glass houses protected vegetables from snow and frost during winters. On my return to India, we constructed a cheap glass house at Kosbad Institute and prepared vegetable seedlings of tomato and chillies during the rainy season. The seedlings were protected from torrential rains. Many farmers in our area constructed plastic houses for preparation of vegetable seedlings during rainy days.

Subsequently during 1964, I visited Holland where I studied reclamation of salt lands. The Dutch scientists gave great thrust on laying of drainage which helped to leech salts from the soil. They also developed some grasses which were salt tolerant. On my return to India I tried these experiments on the lands adjoining the sea coast. We found that the proper drainage helped to leech the salts. We planted paragrass in these reclaimed lands and found that it grew well. This grass was useful for dairy animals. I also introduced salt tolerant trees like *Prosopis julifera* which grew very well in reclaimed soils. I found that reclaimed soils could be utilised for growing fuelwood plantations and fodder cultivation.

I was invited by the United States Government in 1967 to train the American Peace Corps in Applied Nutrition Programme. Accordingly I went to Brattleboro in Vermont State. I visited one farmer who had developed a forest on hundred acres of land. He told me that the Vermont is a hilly State from where number of rivers originated. It is therefore necessary to protect the catchment areas. As a citizen of this State he made humble effort to protect it by afforestation. I was greatly impressed to see the environmental conscienceness among the farmers. On my return to India we undertook an afforestation project on a hilly land of Kosbad Institute. This project was shown to Sarpanchas (Chairman) of the Village Panchayats (Village Councils) who were coming for training at the Panchayati Raj Training Centre. One of the Sarpanchas undertook plantation of casurina trees on the common lands adjoining the sea coast. These trees served as shelter belts and reduced soil erosion caused by wind and sea. Such environmental conscienceness needs to be introduced among farmers and functionaries of Village Panchayats.

I visited Australia in 1971 at the invitation of Community Aid Abroad. During this visit I had an opportunity to study their pastures and fodder development. I found that productivity of animals in Australia had greatly improved due to development of nutritious fodders. While returning from Australia I carried seeds of leguminous fodders which could be grown under rainfed and dry conditions. I conducted experiments for introduction of these fodders at Kosbad Institute and found that *Stylosanthes hamata, Stylosanthes scabra* and siratro were the three fodders suitable both under rainfed or dry conditions. I tried to introduce these fodders all over the country.

The Indian Council of Agricultural Research deputed me to England under the Colombo Plan to study the British Agricultural Education System in 1976. The Agricultural colleges in England were run on commercial lines. I have tried to bring this concept in the horticultural division of the Kosbad Institute. We started a horticultural nursery for supply of quality plants to farmers. There was good demand for such plants and the nursery could be run on commercial lines. I studied rotational grazing of sheep in pastures. They erected electrical fence for that purpose. On my return to Kosbad Institute we divided our grassland into four units by putting a simple cactus fence. The sheep were allowed to graze according to rotation. We found that such practice was possible under Indian conditions. I attended a short course on "Tree Surgery" at Lancashire College of Agriculture. I learned the technique of shaping the trees by proper pruning. Every tree which was planted along the road side was given a shape which would beautify the avenue. Appropriate tools had been developed for that purpose. On my return to Kosbad Institute I requested Shri Mohan Parikh of Yantra Vidyalaya at Bardoli to develop such tools for pruning the trees. Shri Parikh had worked extensively for development of agricultural tools. He immediately prepared appropriate tools for pruning of trees. There were large number of Ashoka trees on both the side of road leading to Kosbad Institute. These trees were not of a good shape. We pruned these trees and gave them proper shape. The whole avenue looked beautiful. I feel that if a course of "Tree Surgery" is offered in Krishi Vigyan Kendras it would help in beautification of the road side avenues.

I was invited by Government of Israel in 1978 to study agriculture in that country. I studied 'Moshav' which was system of cooperative farming. The farmers were doing various enterprises like poultry, dairy, horticulture on individual basis. However, the supply of inputs and marketing was done on cooperative basis. I also studied the 'Kibbutz' which was a system of collective farming. The production and marketing of produce was done collectively in this system. I found that due to cooperative or collective systems of farming the farmers in Israel were highly organized. This facilitated the creation of infrastructure like drip and sprinkler irrigation, marketing, processing and exports of vegetables, fruits, flowers, milk and meat. It also helped to achieve high productivity of crops, animals and birds. On my return to India I undertook a programme to establish cropwise organizations. Thane district has a large area under chikoo (sapota) cultivation. We established a 'Chikoo Growers Organization'. The programmes for improvement in productivity, marketing and exports were undertaken under the auspices of this organization. I have observed that for creating infrastructure in agriculture cropwise organisations of farmers would be very useful in India.

I stayed with one farmer in Moshav who had olive cultivation. He showed me the technique of girdling on olive branches. This technique consisted of preparing a circular ring by removing the bark of the branch without disturbing wood. This was achieved by a sharp knife. Due to girdling the productivity of olives had greatly increased. On my return I experimented this technique on litchi. I found that by adopting this technique the flowering in litchi could be regulated. I arranged a demonstration of girdling. Shri Ramchandra Churi, a progressive farmer from Borigaon village adopted this technique and improved productivity of his litchi plants.

In Israel tomato and caspsicum were cultivated on raised beds. Due to the loose soil on raised beds there is good aeration in the soil. That was one of the reasons for high productivity. On my return to Kosbad Institute I laid an experiment of planting tomato and capsicum on raised beds. Irrigation was given through a channel in between two raised beds. We found that there was considerable improvement in the productivity of these two vegetables. We held a demonstration of this method. The farmers in the region were following the practice of flat beds for planting tomatoes and capsicum. Due to direct watering in flat beds a crust was formed on the upper surface of soil. This crust prevented aeration in the soil, and resulted into low yields. After seeing the demonstration, the farmers started planting tomatoes and capsicum on raised beds.

Since Israel is short of land there are no pastures for grazing cows. Some open land is provided around the cattlesheds for movement of cows. I was told by the scientists that in tropical countries stall feeding could be practised successfully if there were no grazing lands. However, more feeds had to be given in place of fodders. The productivity of milch animals had been greatly improved in Israel by following this practice. Since Israel was short of feed they imported large quantities of feeds from other countries. I explained to the Israeli scientists that in country like India where grain was in great demand for human consumption it would not be possible to give more feeds to the milch animals. They advised me to feed animals on fodder having high protein content. On my return I followed the practice of stall feeding of cows. We fed them with fodder of *Stylosanthes hamata* or *Stylosanthes scabra* which had more protein content. By following this practice the productivity of milch animals was greatly enhanced. In India the area under grazing land is decreasing day by day due to pressure on land for growing food crops. In such cases stall feeding will have to be adopted in future. For that purpose programmes will have to be undertaken to grow leguminous fodders with high protein content.

The vegetables, fruits and flowers from Israel are exported to Europe and America. Farmers from Moshav and Kibbutz were given continuous training in production and post harvest technology required for exports. The extension agency in Israel gave opportunity to the farmers to visit and study the foreign markets. This created export consciousness among the farmers. In the State of Maharashtra such export consciousness had been created among farmers through the Grape Growers Association. These farmers were now exporting their grapes to Europe, East and West Asian countries. Growers Associations would play an important role in promotion of agricultural exports in India.

Israel had adopted a technology of drip and sprinkler irrigation, in order to save its meagre water resources. In addition it improved productivity of crops. On my return I set up a sprinkler irrigation in horticultural nursery at the Kosbad Institute. This was appreciated by the nursery men in surrounding areas and all of them set up such sprinkler units. I have also tried to introduce drip irrigation method among the farmers who grow vegetables. The farmers in Vangaon area have set up drip irrigation systems for growing tomatoes and capsicum. They have achieved very high yields of these two crops due to drip irrigation systems. Some of the farmers who were short of water have set up drip irrigation for their chikoo (sapota) plantations also.

The International Institute of Rural Reconstruction in Philippines invited me to attend a course on 'International Leadership Training' in 1980. During this visit I studied mixed farming systems in that country. The farmers planted coconuts, coffee and rice as mixed crops. This provided continuous cash flow to the farmers. On my return I restructured rice fields and constructed bunds at 20 meter distance. Coconuts were planted on these bunds and rice was taken as an intercrop. We found that mixed farming of coconuts and rice was suitable under the agro-climatic conditions prevailing in the Konkan region. Many farmers in the area have adopted this pattern.

I visited a farmer in Philippines who raised fish in his paddy field. He prepared a trench on one side of the field. After the harvest of rice crop the fish from the field took shelter in the trench. This idea of rearing fish in a small trench was very novel. On my return we practised this method at the Kosbad Institute and it became popular among the tribal farmers.

I visited the International Rice Research Institute at Los Banos in Philippines. This Institute had evolved high yielding varieties of rice like I.R.-8. The scientists at the Institute showed me the technique of growing blue green algae which could be utilized as manure for rice crop. We too started following this technique at the Kosbad Institute. The scientists also showed me a 'rice garden' where rice crop was grown all the year round. We adopted this practice and found that by using early varieties three crops of rice could be raised during the year. The scientists were evolving appropriate farming systems for rice areas of different countries. In one such system a line of maize was planted in between six lines of rice. In another experiment cabbage was planted. We followed this practice of growing vegetables in upland rice and it worked well.

The International Institute of Rural Reconstruction was founded by Dr. Y.C. James Yen. He also established the Philippines Rural Reconstruction Movement. Dr. James Yen first started rural reconstruction experiment in Tingshen in China. The idea of barefoot doctors was evolved by Dr. Chen who was colleague of Dr. James Yen. After coming to Philippines, Dr. James Yen evolved an idea of preparing *barangay* (village) scholars on the lines of barefoot doctors. For that purpose he established 'People's Schools'. Rural youths were given practical training in agriculture, animal husbundry, health etc. They were also given leadership training. When we went to villages one *barangay* scholar was training the farmers in growing high yielding varieties of rice, another scholar was vaccinating cattle, a third lady scholar was training the women in preparation of coconut nursery. A second lady scholar was working in the area of rural health and had instruments for measuring blood pressure, thermometer, weighing balance etc. She kept medicines for common ailments. I found a combination of professional knowledge combined with a spirit of social service among these *barangay* scholars.

Subsequently, I took up a programme of training tribal youths in various skills of agriculture. I also gave them leadership training. These trained youths are now helping the tribals in growing high yielding varieties of rice, vegetable gardening, budding and grafting of fruit trees, setting up nurseries etc. These youths have been very helpful in transfer of technology among the tribal farmers.

Dr. Y.C. James Yen believed in an integrated approach to rural reconstruction, consisting of livelihood, education and culture, health and self government. He developed the following credo for rural reconstruction:

Go to the Peasant people
Live among them
Learn from them
Plan for them
Work with them
Start with what they know
Build with what they have
Teach by showing
Learn by doing
Not a show case
But a pattern
Not odds and ends
But a system
Not a piecemeal
But integrated approach
Not to Confirm
But to transform
Not relief
But release

I introduced the concept of integrated approach of rural reconstruction among the functionaries of Village Panchayats (Village Councils) who were coming for training at the Panchayat Raj Training Centre at the Kosbad Institute. We trained them in the preparation of village development plans using such an integrated approach.

I got the opportunity to study agriculture and rural development practices in various countries. On my return to India I have persistently tried to replicate the new agricultural technologies imbibed by me. I have also tried to introduce the organizational methods in cooperatives or rural development programmes wherever it was possible. I strongly feel that such introductions are possible if suitable modifications are made suiting the Indian conditions.

13

CONTACT WITH NATIONAL LEADERS

Visits of National Leaders

National leaders from all over India visited Kosbad Institute to see our work for tribal development. Shrimati Indira Gandhi, former Prime Minister of India, visited Kosbad Institute in 1958. We arranged an exhibition of flowers grown in the region, and the farmers brought their exhibits. Shri Merwan Khodadad Irani and Shri Jehangir Irani who acquired skills in growing cut flowers explained to Smt. Gandhi how flowers were grown as inter-crops in Chikoo plantations. They imported the bulbs of *Gladiola flowers* from Netherlands to maintain quality. They also told her that the farmers ran a floriculture shop in the Taj Mahal Hotel at Bombay. Shrimati Gandhi was happy to see the development of floriculture in the region.

Shri Merwan Irani carried various flowers from his garden for Smt. Gandhi whenever he visited Delhi. She also often invited Shri Irani for dinner so that she could get to hear his experiences in agriculture. Once he told her about role of earthworms in agriculture. Shri Irani used organic manures in chikoo gardens. This had led to the growth of earthworms. He further told her that the presence of earthworms had enhanced his soil's fertility. Smt. Indira Gandhi was of firm conviction that agricultural development should be done keeping in view environmental balance. She therefore, appreciated efforts of Shri Irani in promoting growth of earthworms. She even quoted this example in one of her public speeches. I met Shrimati Indira Gandhi whenever I went to Delhi and continued to inform her about new developments in floriculture. She expressed her appreciation for Acharya Bhise and Shrimati Tarabai Modak for their work for the upliftment of tribals.

The member of Parliament of our district Shri Bhausaheb Dhamankar, alongwith two guests visited my house one morning. He told me that they had come by the Frontier Mail from Delhi and would be visiting the Nuclear Power Station at Tarapur at noon. He therefore, decided to show them our work at Kosbad Institute in the morning. These guests were Shri Vishwanath Pratap Singh and Shri Chintamani Panigrahi who were also Members of Parliament at that time.

We first took them to the *Balwadi* and creche centre created for the tribal children. My wife Meena who organized this *Balwadi* explained to them that the working mothers brought their children and infants in morning. While the children attended the *Balwadi*, the infants were kept in the creche. The mothers then went to the nearby fruit orchards for labour. She also told them that a mid day meal was arranged for children as many a times they could not have food in their homes. We showed them the Nutrition Garden attached to the *Balwadi* where vegetables were grown for midday meals for the children. We told them that such integrated approach of creche, *Balwadi* and nutrition gardens is essential for imparting education to children from weaker sections.

We also showed them the work experience projects of tribal student in secondary schools which comprised vegetable gardening, poultry keeping, calf rearing, grafting and budding. The students told them that the skills that they acquired through these projects could be applied when they went back to their villages. They transformed the local mango trees into table varieties like Alphanso, *Payari* and *Kesar* by grafting. They also transformed the local *ber* trees into improved varieties like *Gola* and *Umran* by budding techniques. The girl students told them that they are rearing improved breeds of poultry which lay more eggs. The visitors were very happy to see the skill training being put to practice. They were of the opinion that this was a real life education.

The visitors were taken to the Ashram school too. The headmistress of the Ashram School was Shrimati Kalavati Pardhi. Kalavati belonged to Scheduled Tribe. She received her primary and secondary education at Kosbad Institute and then completed her teacher's education in Vikasvadi founded by Shrimati Tarabai Modak. Kalavati developed the idealism of social service practiced at the Kosbad Institute and decided to devote herself to the cause of educational upliftment of tribal children. Her husband Shri Sakharam Pardhi also trained at the Agricultural Institute and became a teacher in the Ashram School. Shri Vishwanath Pratap Singh was so impressed to see the devotion of this tribal couple that he told me to send one such couple to one of his institutions' in Uttar Pradesh.

We showed to the visitors fruit orchards of tribals in the surrounding villages. The tribal farmers told them that for generations they were working as labourers but because of the support offered by the Kosbad

Institute they were able to develop their own orchards. Shri Vishwanath Pratap Singh commented that the weaker sections needed such support services to lead a life with self respect.

Shri Singh wrote me after returning to Delhi and complimented us for our work at the Kosbad Institute. Shri Vishwanath Pratap Singh became the Prime Minister of India in 1989 and Shri Chitamani Panigrahi became the Governor of Manipur.

Contact with National Leaders

We conducted study tours of trainees from Kosbad Institute. Most of these trainees belonged to Scheduled Tribes. They were taken to see various agricultural development projects in the country. While the trainees were in Delhi we took them to meet Dr. Rajendra Prasad, President of India. Dr. Rajendra Prasad told them that when he visited Bordi for the All India Tribal Welfare Conference the idea of starting an agricultural training for tribal youth was placed before him by Acharya Bhise. He advised the tribal youths to propogate new technology in agriculture among the tribal farmers of their villages. He told them that the hunger and malnutrition prevailing among the tribals could be removed by adopting improved methods in agriculture.

We took the trainees to meet Pandit Jawahar Lal Nehru, Prime Minister of India. He was also very happy that the tribal youths were given training in improved methods of agriculture. He told them that whenever he visited the tribal areas in the country he found that there was widespread problem of malnutrition among the tribal people. He advised the tribal youths to utilise their knowledge for growing food crops, vegetables and fruit trees. By growing these crops they could get a balanced diet which would help to overcome the problem of malnutrition. Pandit Nehru was very happy when we told him that the theme of this training was based on the principle of 'learning by doing'. The trainees were greatly inspired by the advice of the President and Prime Minister of India. They adopted the new technology in agriculture not only on their own lands but in the adjoining lands too.

Shri Morarji Desai, Prime Minister of India was taking keen interest in the tribal welfare work. He was President of the Adimjati Sewak Sangh which was founded by Late Thakkar Bappa. I used to meet him very often and conveyed the account of work done at the Kosbad Institute. He advised me to organise training for tribal youths from all over India. The Adimjati Sewak Sangh thus made the arrangements to depute trainees from all the States. We organized training in cultivation of high yielding varieties of cereal crops, pulses and oilseeds. We also gave them training in agro-forestry and horticulture. The trainees were taken on the study tour to see the Forest Labour Cooperatives, Ashram Schools and Institutions run by non-governmental voluntary organizations. The trainees after the completion of the course were satisfied as they become well versed in skills required in agriculture as also gained an insight regarding the tribal development work. They also expressed that such courses are very useful for bringing national integration.

The Chief Minister of Maharashtra Shri Sharad Pawar sent me a message in 1979 that I should come to Dabchari Dairy Development Project which was about 30 kms. from the Kosbad Institute. The Project Officer told me that Shri Pawar had gone to Surat and will be returning in the evening. Accordingly Shri Pawar came with Shri Chandrasekhar at in the evening. Shri Chandrasekhar was the President of Janata Party that time. Shri Pawar introduced me to Shri Chandrasekhar and requested me to tell about the activities of the Kosbad Institute. I told him about the horticultural development work which we were doing among the tribals of Thane district. He was very happy to learn that the tribals who were working as labourers for generations could now develop their own orchards of chikoo, mango and coconuts. I told him that these orchards created gainful employment on their lands and the tribals did not have to migrate to the urban areas in search of employment. Shri Chandrasekhar expressed the desire to visit Kosbad Institute in the near future. However, he could not do so. When he became Prime Minister of India in 1990 I met him in Mani Bhawan at Bombay. He remembered about the tribal upliftment work which we were doing at Kosbad Institute.

14

VISIONARIES

Swami Anand

I was fortunate to receive valuable guidance in my life from Swami Anand. During his early days Swami Anand stayed at Bombay. One day he met a *Sadhu* (monk) to whom he said that he wanted to see God. The *Sadhu* replied that he would certainly see God if he accompanied him to Nasik. However, his desire could not be fulfilled and hence, Swami Anand went to the Himalayas to stay with *Sadhus* for several years. He studied religious scriptures and Indian Philosophy. Thereafter, he became a monk of Rama Krishna Mission.

When Mahatma Gandhi came to India from South Africa, Swami Anand felt that this great man would alone show the real path to India. He therefore, decided to work with Gandhiji. When Mahatma Gandhi came to know that Swani Anand was experienced in managing printing press and publications, he assigned him the responsibility of management of the Navjivan Press at Ahmedabad . Gandhiji published *Navjivan* and *Young India* periodicals from this press. Mahatma Gandhi was of the firm view that the periodicals be published in time and be devoid of any printing errors. Swami Anand followed the directions of Gandhiji and committed himself to the Navjivan Press. The Congress session was held at Ahmedabad, and everybody was surprised on Swami Anand's absence. At this point Sardar Patel replied that Swami Anand would be at the Navjivan Press. Swami Anand told me that he learned discipline in life from Mahatma Gandhi. When he was given the responsibility of managing the Navjivan Press he never left the press. He followed the instructions of Gandhiji to see that his periodicals were issued in time and did not have any printing errors. Swamiji advised me that I too should maintain discipline in my work at Kosbad Institute.

He also advised me that instead of spreading our activities in a larger area we should try to serve the poor people in nearby villages. He told me that at one of the Conferences of the Christian Missionaries in India, Gandhiji advised them to follow the spirit of Swadeshi. He defined Swadeshi thus : "Swadeshi is that spirit in us which restricts us to the service of immediate neighbour to the exclusion of the more remote". We followed the advice of Swami Anand and restricted our rural reconstruction work in the neighbouring tribal villages.

Swami Anand told me the example of Bardoli Satyagrah (Non-Cooperation Movement) during 1928. The farmers from Bardoli Taluka approached Gandhiji and told him that the British Government had increased land revenue which they were unable to pay. They therefore, wanted to organise a movement against this injustice. Gandhiji requested Shri Vallabhbhai Patel to lead this movement. Gandhiji desired that Swami Anand and Shri Jugatram Dave assist Shri Vallabhbhai Patel in this movement. Shri Vallabhbhai Patel concentrated his energy on this movement. He organized meetings of the farmers in every village and appealed to them not to pay the increased land revenue. The farmers responded to his appeal, and not a single farmer paid the land revenue. Women also joined the movement. The British Government attached the properties of the farmers, and declared auctions, but nobody came forward to purchase the auctioned material. This resulted in a strong unity among the farmers. Shri Jugatram Dave used to take down the speeches of Shri Vallabhbhai Patel and bring them for printing to Surat. Many a times during the rainy season, rivers and streams were in spate. In such conditions also Shri Jugatram came regularly to Surat. Swami Anand who was staying in Surat printed the events of the Bardoli struggle in the form of a newsletter. This newsletter was so popular that people had to stand in a queue to purchase it. The newsletter was issued regularly without a break. When the struggle of Bardoli had been going on continuously for several months, there was a debate in British Parliament in which one member of Parliament remarked that Government should make some settlement in land revenue issue, otherwise the whole of India would be Bardolized. Accordingly the British Government had to withdraw the increased revenue and make the settlement. This was a great victory for Bardoli farmers. The farmers of Bardoli gave the title of 'Sardar' to Shri Vallabhbhai Patel. From that time onwards he is known as Sardar Vallabhbhai Patel to the people of India. Swami Anand told me that during Bardoli Satyagrah they concentrated on one issue of 'Non-Cooperation' and therefore, they were successful in this struggle. He advised me to follow this approach in rural reconstruction work.

Some landlords in nearby villages used coercive methods to evict the tribals from their lands. Such tribals came to us for help. Swami Anand told us that whenever such evictions took place we should leave other work and fight against injustice. Even though we were single handed we should continue our fight. He quoted the example of Guru Govind Singh who fought bravely with the Moghuls with only five disciples. He described his disciples as follows :

"I would kill hawk with sparrows
I would kill tiger with cat
I would make my one disciple to fight with
One hundred twenty thousand Moghuls.
Then only I would call myself Guru Govind"

Swami Anand used to quote an example of the American Scientist, Prof. George Washington Carver. Prof. Carver was born in a slave negro family in 1860. He was an intelligent student and graduated from Iowa State College of Agriculture at Ames. Thereafter, he was offered a position at the college as assistant botanist, specializing in hybridizing fruits. Not only was he the first to graduate from College but was also the first black to serve on the faculty. In 1896 Prof. Carver was asked by Dr. Booker T. Washington to join the staff of Tuskegee Institute in the State of Alabama. After his acceptance, he immediately started an agricultural experiment station. His objective was to build up wastelands of the south debilitated by the one-crop system (cotton being the single crop grown) and find new commercial uses for other native crops. Though his basic intent was to aid impoverished blacks the white farmers were in almost as bad a state. Prof. Carver aided both with his knowledge.

Prof. Carver developed 118 products from sweet potatoes which could be used as food substitutes. Using groundnut (peanut) he created more than 300 products. Many industrialists approached Prof. Carver to accept a job in their industries on a high salary. However, he politely declined these offers saying that his first duty was to serve the poor black farmers. Swami Anand told me that one has to choose either career or commitment in his life. Prof. Carver chose commitment in his life. Swami Anand told us that India needed Scientists like Prof. George Washington Carver who would commit themselves to solve the problems of poor and resourceless farmers.

Swami Anand told me that Prof. Carver used to write letters to Mahatma Gandhi admiring his role in promoting truth and non-violence in politics. Prof. Carver further wrote that whatever reserches he made in science were for the welfare of humanity. Gandhiji admired the contributions made by Prof. Carver.

Swami Anand believed that the scientists in our country needed to choose one subject and conduct research on it for life long. He quoted the example of Dr. Robert Openheimer who made great contribution in physics. Swami Anand, who wrote the biography of Dr. Openheimer, advised me to concentrate on research of food crops as millions of people in our country suffered from hunger and malnutrition. He was very happy when we introduced high yielding varieties of rice and wheat. He was pleased to see the three crop cycle of rice, wheat and *mung* which we evolved at the Kosbad Institute.

According to Swami Anand, the agricultural scientists and extension workers should assume the role of lotus. The lotus stays at one place, but its fragrance and nector attracts insects and bees from long distances. We followed his advice and found that people from India and abroad visited the Kosbad Institute to see the agricultural research and rural reconstruction work.

Swami Anand told me that cow was an integral part of Indian agriculture. Cow's milk is nutritious food for human beings and cow dung is food for the soil. Therefore, we call cow as mother. Cow also provided bullocks which help in tilling the land. Swami Anand told me one interesting story. Dr. S. Higginbottom, an English Scientist worked for several years at the Agricultural Institute in Allahabad. A demonstration of ploughing the land with tractor was arranged in a nearby village by the manufacturing firm. Dr. Higginbottom was asked to give his comments about this demonstration. Dr. Higginbottom said "the tractor has done good work of tilling the soil, but India needs a tractor which would till the soil and give dung" Dr. Higginbottom was a great visionary and knew that the agricultural productivity in India would only be improved only if cattle dung manure was applied to soils.

Swami Anand advised me to keep a herd of cows in the Kosbad Institute. He specially took me to 'Lokbharti Institute' at Sanosara in Saurashtra where a good herd of Gir Cows was maintained. The workers of Lokbharti helped us to secure Gir Cows for Kosbad Institute. Since then we started maintaining a herd of cows. He took great interest in fodder development work at the Kosbad Institute. He was very happy to know that I was able to introduce leguminous fodders like stylosanthes and siratro from Australia. He suggested vernacular names to these fodders as follows:

Stylosanthes	Shravan Ghas
Siratro	Sita Ghas
Desmanthus	Dashrath Ghas

These names have become popular among the farmers.

Shri Narayan Desai used to come very often to meet Swami Anand. Narayan was a son of late Shri Mahadeo Desai who was Secretary to Mahatma Gandhi. Narayan Desai was a sarvodaya worker and used to tell his experiences about rural reconstruction work done by sarvodaya workers in India. During the Centenary Celebrations of Mahatma Gandhi, a delegation of followers of Rev. Martin Luther King Jr. was invited to India. Their programme included attending seminars and meetings in Delhi. They told Shri Narayan Desai "we came with high expectations to see Mahatma Gandhi's teachings in action. However, we do not see Gandhi anywhere." Shri Narayan Desai took them to Vedhachhi, a remote village in Surat District of Gujarat. They visited a camp where more than five thousand tribal children were spinning on *charkha* (Spinning Wheel). On seeing this site, they remarked "Now we see Gandhi." Shri Narayan Desai explained to them that Mahatma Gandhi created contructive workers like Shri Jugatram Dave who committed their life for rural upliftment. Shri Jugatram was present at that time.

Shri Jaiprakash Narayan organized *Shanti Sena* (Peace brigade) in North East Frontier Agency (Now known as Arunachal Pradesh) after the Chinese war in 1964. The workers of *Shanti Sena* stayed in villages near the India-China border and did rural reconstruction work. Shri Narayan Desai was given the responsibility of training these workers. Shri Desai desired that I should come there to impart training in agriculture. Shri Desai had organized a training camp for these workers near Zero. I stayed there for a week and gave them training in agriculture. I got an opportunity to study the agro-climatic conditions of NEFA and found that the region has great potential for horticultural development.

Swami Anand used to tell me that our agriculture has been based on thousands of years of experiences. Even though our farmers may not have gone to school, yet they acquired the wisdom of practicing agriculture. He narrated the experiences of Konkani Muslim farmers in Bhiwandi Taluka of Thane district. As soon as first shower of rains came, these farmers started running to their fields. They were keeping basket of seeds on their heads. While running they were singing a folk song loudly which said "Let us run to our rice fields. Even though our father is dead, we should cover the dead body with cloth. The seed which is in our hand is more effective than which is in the basket and the seed which is in the palm is more effective than that which is in the hand." This means that after the first shower of rains, every second is important in sowing of rice seeds. This shows that our farmers inherited knowledge of timely sowing of seeds.

Swami Anand had been staying at Kausani in Almora district of Uttar Pradesh for many years. The farmers in that area planted rice. Water from natural streams was diverted in such a way that it used to pass from one field to other. Because of the continuous flow of water, rice crops got fresh oxygen thereby increasing yields. Swami Anand used to say that our farmers had inherited the knowledge of water management for generations.

Great men from all over India used to come to Kosbad Hill to meet Swami Anand. Prominent among them were Shri Ravishankar Maharaj, Smt. Mani Ben Patel, Dr. Uma Shankar Joshi, Shri U.N. Dheber, Shri Gangadhar Rao Deshpande, Shri Shankarrao Dev, Shri Dada Dharmadhikari, Shri R.K. Patil, Shri Shriman Narayan and Shri Vaikunth Lal Mehta. Swami Anand showed them the work of the Kosbad Institute. I got an opportunity to get guidance from such great persons in my life.

Swami Anand stayed at the Kosbad Institute for about 19 years between 1957 and 1976. He was a person of great conviction and believed that voluntary workers like me instead of spreading activities of rural reconstruction in large areas should select a small area and do concentrated work. Such work would create impact on the community and provide guidance to others. I followed his advise and did my work in that direction.

Shri Shivajirao Bhave

Shri Shivajirao Bhave younger brother of Shri Vinobha Bhave often used to come to Kosbad Institute to meet Swami Anand. He was a person of great knowledge and wisdom. I used to show him various activities of our Institute in Rural Reconstruction.

Shri Bhave told me that there existed four cultures in the World. In Africa, there was *Bal Sanskriti* (child culture). People fought amongst themselves and the nations also had conflicts with each other. There was thus no peace and stability in the continent. In middle East or West Asia, there was *Maidani Sanskriti* (desert culture). People invaded those parts of the desert where there was prosperity and amassed wealth by looting. In Western countries, there is a *Prayogic Sanskriti* (culture of research). There is a severe winter for five or six months and due to snow people have to confine themselves in homes or buildings. So, they constructed buildings to house educational institutions and research laboratories. This enabled them to produce good scientists. They made great discoveries and inventions. In India, there is *Adhyatmic Sanskriti* (spiritual culture). This culture was created by our ancient saints and our society has preserved this culture for thousand of years. The kings and princes too respected this culture. Chatrapati Shivaji visited. Saint Ramdas and Tukaram to follow their spiritual advice. Shri Shivajirao advised me that while doing rural reconstruction work we too must see that this culture is preserved.

I asked him why the universities in the West such as Cambridge, Oxford, Harvard or Princeton have maintained their educational standards for hundred of years. While in India, we have not succeeded in maintaining such educational standards. He was of the opinion that big institutions would not be suitable for our country. In ancient times, our teacher saints used to stay away from towns. They had their Ashrams in the deep forests. These Ashrams were like small universities. Very few students were admitted at a time and so there was a close interaction of knowledge between teachers and pupils. We have broken this tradition and started big institutions on Western patterns and so the educational level in our case has deteriorated.

Shri Shivajirao asked me, "how many teachers you have in your Institute?" I replied, "about one hundred." He further asked me, "Are you in a position to have dialogue with each of them?" Is it possible for you to know their thinking? I replied that it was not possible for me to meet them for several days. He further told, "When you are not able to meet even your staff then how is it possible to have close interaction with your students?" He said this is a mechanical process of bookish education. Real education is that which prepares students to lead an ideal life. Such an education is possible only through close interaction between teachers and students. I asked him, "What should we do to overcome this?" He advised me to make the institution small. He told me to encourage my staff to start small institutions in their areas and admit few students with whom they would have closer interaction. He narrated his own example. He wrote concordence on "Dnyaneshwari" which took a number of years to complete. He required several references for this purpose. He entrusted the work to Shri Haribhau who was the son of a farmer. Haribhau did not have any higher education but due to his close association with Shri Shivajirao Bhave, he became proficient in reference work.

Shri Shivajirao Bhave told me that your institution has been started with an objective of doing rural reconstruction work. This would require very committed workers. Such workers could only be prepared if there exists a close interaction between the head of the institution and the workers. He further said that every inmate in the Ashram of Mahatma

Gandhi got an opportunity to discuss his or her ideas with Gandhiji and thus they received his personal guidance in rural reconstruction work. I was convinced by the advice of Shri Shivajirao Bhave that we too should have small institutions in our country. Such institutions alone could produce committed workers for rural reconstruction.

Acharya Vinobha Bhave

I got the opportunity to meet Vinobhaji thrice at his Paunar Ashram near Wardha. He gave me valuable guidance which had considerable bearing on my subsequent actions. Vinobhaji could not hear and so his secretary used to write the questions which we wanted to ask. Vinobhaji used to read those questions and reply them.

I asked him how should we utilize land in India? He replied "wherever there is vacant land utilize it for growing food crops. Food crops should be grown on both the sides of railways and roads. In future the land would fall so short that only Baba's *Padayatra* (walk) or Kissinger's *hawai yatra* (air travel) would remain." I told Vinobhaji that we have been trying to introduce improved methods in agriculture among tribals for last thirty years. However, many tribals have not yet adopted the new methods. What was the reason for this? Vinobhaji asked me, "while moving among the tribal how many times you have joined them for lunch in their homes?" I admitted that I did not do so. He told me, " whenever you move among the tribals join them for meals in their huts. You could carry some rice and *dal* (pulses) with you so that it would not be a burden on them. Take only vegetarian meals." He implied that this would enable the tribals to identify with me and hence appreciate and accept the programmes.

I told Vinobhaji that Swami Anand advised me to work life long at the Kosbad Institute. Vinobhaji replied, "What Swami Anand says is perfectly right. I would further add that you should stay at the Kosbad Institute even after your death."

Once, I attended a prayer meeting of Vinobhaji during his *Bhoodan Yatra* (land grant mission) in Maharashtra. Vinobhaji said that he was going to Wardha from Paunar in early morning. He saw a farmer standing in his *Jowar* (sorghum) field. The farmer had a sling in his hand with which rubble could be thrown to scare the sparrows. A large number of sparrows were eating the grains while the farmer was simply watching quietly without scaring away the birds. Vinobhaji asked the farmer why was he not scaring the sparrows? The farmer replied that it was *Ramprahar* (the hour of *Ram*; viz. dawn) and so the birds have a right on the grains. After sunrise we would scare them. Vinobhaji said in the prayer meeting that our farmers have a tradition of farming for fourty centuries. They have inherited wisdom that all living creatures on the earth play some useful role. They know that the sparrows are very helpful in protecting food crops from insects, pests and diseases.

I was greatly inspired by the valuable advice of Vinobhaji. It provided me guidance in my work of rural reconstruction.

Acharya Bhise

One of the great persons who shaped my life was Acharya Bhise. He was born in 1894 in a remote village 'Sai' in the Raigad district. His parents were very poor and were unable to support him for his high school education. However, Shri Kharkar a great philanthropist from Thane took the responsibility of his education. He was a bright student in his school career. After passing his matriculation examination he entered Wilson College at Bombay. After receiving his B.A. Degree, he decided to devote himself to the cause of education. He also decided to remain a bachelor so that he could devote full time to his mission. His teacher Shri T.A. Kulkarni had founded the Gokhale Education Society. Shri Bhise joined this Society. A new high school was started at Bordi, a village in Thane district under the auspices of this Society. Shri Kulkarni wanted Shri Bhise to accept the post of Principal of this School. However, Shri Bhise did not appreciate this idea as he was brought up in the city of Bombay. He told Shri Kulkarni that he would rather work in the labour areas of Bombay. Shri Kulkarni reasoned that there were enough educated people working in cities, but villages need educated youths with ideals. However, the advice did not convince Acharya Bhise to go to Bordi. A few days later, he went to his maternal Uncle's place at 'Mahad' in Raigad district. While going around the villages he came across a grave in a lonely place. On enquiry he was informed that one missionary doctor from U.S.A. came to this village and served life-long for the welfare of the poor people. He died here and was laid to rest in the land. The Acharya was moved to learn that a missionary doctor had come 10,000 miles to devote his life for the service of the poor, while he was not prepared to go to Bordi which was only eighty miles from Bombay. He felt very sorry for having ignored the suggestion of Shri Kulkarni for going to Bordi. He immediately returned to Bombay and expressed his regret to Shri Kulkarni and agreed to go to Bordi. Accordingly he came to Bordi in 1920 and served the villagers and tribals for more than fifty years.

I was a student in S.P.H. School at Bordi between 1937 and 1944. Acharya Bhise was the Principal of the High School. This was the period of the British rule. Struggle for independence was going on under the leadership of Mahatma Gandhi. Acharya Bhise participated in 'Individual Disobedience Movement' in 1939 and 'Quit India Movement' in 1942, and was imprisoned both times. Acharya Bhise had a team of very devoted teachers in his school. Many of them actively participated in the freedom struggle and had undergone imprisonment.

Sane Guruji, a saintly Poet and author of Marathi literature was staying in Bordi High School. His brother, Shri Purushottam Sane was the Head Master of the School. Sane Guruji gave sermons about India's freedom movement in the prayer meetings. This has instilled a spirit of nationalism in my life.

Acharya Bhise had started vocational training in agriculture, spinning and weaving, hand-made paper, palm juice (*Neera*) and jaggery (*Gur*) making. We received skill training in these vocations. Since we were working manually in all these vocations, a sense of dignity of labour was created in me from my student days.

I passed my matriculation examination from the University of Bombay in 1944. I sought the guidance of Acharya Bhise about the type of higher education I should take. He told me that the nation needed rural reconstruction workers and that agriculture plays an important role in rural reconstruction. He desired that I should receive agricultural education. I listened to his advice and accordingly joined the college of Agriculture at Poona. On receiving my B.Sc. (Agriculture) degree of the University of Bombay, I joined "the Kosbad Institute" in June 1949 and started my work under the guidance of Acharya Bhise. I was very fortunate that I received his guidance for all my life.

Acharya Bhise used to tell me that "the Kosbad Institute" should not be kept within four walls but its work should be spread in the surrounding tribal areas. I have followed his advice. This has helped me to study the problems of the tribals and seek solutions to them. Due to this approach the Kosbad Institute has become a community centre.

Acharya Bhise used to advise me that I should treat the staff of the Institute as family members and should help them during their difficulties. He narrated one incidence in his respect. Acharya Bhise was in prison during the 'Quit India Movement' in 1942. His family was in a financial crisis. When Thakkar Bapa, founder of Bharatiya Adimjati Sewak Sangh (All India Tribal Service Organization) came to know this, he immediately sent one representative to find out how much financial help would be necessary. Meanwhile, the family members of Acharya Bhise had made arrangements to resolve the financial difficulties, and so there was no need to seek assistance from Thakkar Bapa at that time. The incidence shows that whenever our colleague is in difficulty he should immediately be extended help.

Acharya Bhise used to tell me that since the Kosbad Institute is situated in the tribal area its role should not be simply one of imparting training but also of human service. Acharya quoted the example of Father Demiyan who served lepers in Hawaii. He trained the lepers to do farming. However, he himself got infected by leprosy but without being nervous he continued his mission. Acharya Bhise expected the staff members in this Institute to work with such missionary spirit.

Acharya Bhise used to tell the story of Eklavya to tribal students. Though he was not admitted to a formal school, Eklavya became one of the best archers in the land, through single-minded devotion to his skill.

Acharya Bhise desired that the tribal students should also acquire such proficiency in agriculture. This would help them to produce enough food and overcome the problems of their people.

Acharya Bhise used to take me to the Conferences of Adimjati Sevak Sangh which were held at Guntur in Andhra Pradesh, Rushabhadev in Rajasthan and Dohad in Gujarat. In these Conferences the tribal problems were discussed at the all India level which helped me to get insight into tribal life across the country.

One such Conference of Bharatiya Adimjati Sevak Sangh was held in Bordi in 1959. The President of India, Dr. Rajendra Prasad and Union Home Minister, Shri Govind Ballabh Pant attended this Conference. The President of Indian National Congress Shri U.N. Dhebhar and Govèrnor of Bombay Shri Prakash were also present.

All the dignitories at the Conference praised the selfless service of Acharya to the cause of tribal welfare. They desired that Acharya Bhise should accept the post of Adviser to Government of India in Tribal Welfare. However, he politely declined the offer stating that he would not leave his tribal welfare work in Thane district. He kept before him the ideal of serving life long at one place.

He was offered Padmabhushan from the Government of India. However, he declined this offer too. Shri D.R. Pradhan, Chief Secretary to Government of Maharashtra specially went to Shri T.A. Kulkarni, Chairman of Gokhale Education Society to persuade Acharya Bhise to accept this honour. Acharya Bhise regarded Shri Kulkarni as his *Guru* (teacher). However, Shri Kulkarni told Shri Pradhan that Acharya Bhise is a man of principle and he has decided not to accept any honour as a principle. Shri Kulkarni therefore expressed his inability to persuade

Shri Bhise.

Acharya Bhise has put in great efforts for the Kosbad Institute. He used to come to Kosbad on foot as there was no road. When the road was constructed he started coming by bus. He rarely used a jeep or any other vehicle. He used to carry a cloth bag filled with papers. Many a times it was too heavy but he never allowed others to carry it. Whenever he came to the Institute he was enquiring about the difficulties of the students and staff. He tried to solve these difficulties by contacting the Ministers and officials in Mantralaya at Bombay. Often, I used to accompany him to Mantralaya. All the Ministers and officials treated him with great respect. Sometimes there were delays due to government procedures. However, the Acharya was never angry. He used to approach even smaller officers for work of the Institute. He never believed that his dignity would be affected by approaching smaller officers. Due to this approach the Kosbad Institute received cooperation from all levels of the Government.

Acharya Bhise never took rest in his life. Sometimes when his blood pressure rose, we would request him to rest. At that time he used to raise his finger to show that he will get rest only in heaven. Acharya Bhise breathed his last on 4th July 1971. He was a real *Karamyogi.* His selfless service to the poor would inspire many generations to come.

Dr. M.S. Swaminathan

Dr. M.S. Swaminathan the architect of the 'Green Revolution' gave me valuable guidance and support in my work at the Kosbad Institute.

I first met Dr. Swaminathan in a wheat field of Indian Agricultural Research Institute at New Delhi. He was showing wheat experiments to Dr. Norman Borlaug. Dr. Borlaug who had received the Nobel Peace Prize for his contribution in wheat research was invited by the Government of India. When Dr. Swaminathan noticed that I was standing alone in the wheat field he came to me and introduced himself. I told him that I was working at Kosbad Institute. I reminded him that Mexican wheat seeds he sent to me were sown in the field and the crop was excellent. Dr. Swaminathan then introduced me to Dr. Norman Borlaug. He invited me for the felicitation function of Dr. Borlaug which was to be held at Vigyan Bhavan in the evening. In my first meeting with Dr. Swaminathan I found that he was a person of great humility and treated rural development workers like me with considerable respect.

I used to convey to him through letters about the research and extension work being undertaken at Kosbad Institute. He immediately replied to my letters. In every letter he wrote about the new researches and developments in agriculture in India and abroad.

Shri M.S. Pawar, Vice Chancellor of Mahatma Phule Agricultural University at Rahuri invited Dr. Swaminathan to Kosbad Institute in 1974. At that time Dr. Swaminathan was Director General, Indian Council of Agricultural Research. We took him to the research section of the Kosbad Institute. Dr. N.R. Bhatt, Dr. R.D. Asana, Dr. S. Solomon, Prof. U.K. Kanitkar and Dr. N.G. Magar were working as emeritus scientists at the Institute. They showed him various research activities on rice, pulses, vegetables and forage crops. He was very happy to see eminent scientists doing research in tribal areas and guiding the post graduate students. Dr. Swaminathan desired that we should strengthen the agricultural research section of our Institute by establishing Plant Science Laboratory. Accordingly, we established a well equipped Plant Science Laboratory at the Institute.

We inaugurated the Fodder Research Centre at the hands of Dr. Swaminathan. He was greatly impressed to see the introduction of various fodders which we brought from Australia. On his return to Delhi he sent us seeds of *Stylosanthes scabra* which he brought from Australia. We introduced this fodder in our grasslands. On his directive the Vice Chancellors of the Agricultural Universities and the Directors of the Central Institute of Indian Council of Agricultural Research visited Kosbad Institute to see the research on forage crops.

Dr. Swaminathan delivered a Convocation Address at the University of Bombay. In his address he mentioned about the work of the Kosbad Institute in development of Subabool *(Leucaena leucocephala).* He said that this tree would cater to the needs of fuelwood, fodder and fertilisers of the rural masses. He requested the staff and the students of the University to visit the Kosbad Institute and see the rural reconstruction work going on there. Dr. Swaminathan very often used to mention about the work of Kosbad Institute in Seminars and Workshops all over India.

During his visit to the Kosbad Institute we showed him the various activities like water conservation, green manures and cultivation of high yielding varieties of rice. We took him to tribal farmers who had introduced high yielding varieties of rice on their farms. The farmers told him that the high yielding varieties have helped them to become self reliant in their food requirement. Before the introduction of these varieties they faced hunger for part of the year. The farmers also showed him their mango, chikoo and coconut orchards. They told him that these orchards helped them to generate gainful employment on their lands.

Dr. Swaminathan was greatly impressed to see the agricultural development work among the tribals. He expressed that the Kosbad

Institute would be the most appropriate place for starting the Krishi Vigyan Kendra (Farms Science Centre). Due to his support the first Krishi Vigyan Kendra in Maharashtra was granted by Indian Council for Agricultural Research in 1976.

Dr. Swaminathan desired that the Krishi Vigyan Kendra at Kosbad should become an ideal centre in the whole of the country. He always used to guide us in that direction. He deputed me to United Kingdom to study the British Agricultural Education System under the Colombo Plan. I found that the sections like crops, dairy, piggery in the Agricultural Colleges were run on commercial lines. This enabled the students to prepare themselves for commercial agriculture. I wrote one Paper on "British Agricultural Education System and its Relevance to the Krishi Vigyan Kendras in India" and sent it to Dr. M.S. Swaminathan. He circulated this paper to Agricultural Universities, and Central Institutes of Indian Council of Agricultural Research.

Dr. Swaminathan encouraged me greatly in my work at the Kosbad Institute. He recommended my name for the Jamnalal Bajaj Award for application of science and technology for rural development. I received this award in 1979, from the President of India, Shri N. Sanjiva Reddy. Due to this award my work was recognized at the national level. Dr. Swaminathan is a great humanitarian who has concern for the upliftment humanitarian of the poor. He has always encouraged and supported individuals like me who are working among the poor people of India.

15

EDUCATIONISTS AND SOCIAL REFORMERS

Shri T.A. Kulkarni

The founder Chairman of Gokhale Education Society Shri T.A. Kulkarni has given me valuable guidance in my rural reconstruction work. Shri Kulkarni belonged to Pali Village in Sudhagad taluka of Raigad district. During his school days he used to go with shepherds to nearby forest areas. They were showing him various plants and vines which were used as green fodder for their animals. From those days he developed interest in fodder plants.

Shri Kulkarni told me that in Konkan the green fodder is available only during rainy season. For the rest eight months which are dry we could utilize leaves of trees as green fodder. He narrated example of *Anjan* tree from Satpuda hills of Jalgaon and Dhule districts. The cattle, sheep and goats in that region are fed on leaves of this tree for generation. The fat content of the milk is increased by feeding these leaves. This tree is drought resistant and could be grown in soil of low fertility. Shri Kulkarni told me to propogate this tree all over Konkan tract. We prepared a plot of *Anjan* trees and arranged demonstration to show that this fodder tree could be grown under agro-climatic conditions of Konkan. Later we prepared seedlings of *Anjan* and distributed them to farmers from Konkan region.

Shri Kulkarni told us that leaves of *Pipal* tree could be utilized for fodder. *Pipal* tree has long life. Presently this tree is planted as shade tree on road side. He told us to collect seedlings from bird droppings. Accordingly we prepared a demonstration plot of *Pipal* trees. Later we conducted an experiment on germination of *Pipal* seeds. We found that when the seeds are kept in incubator and given proper heat treatment the germination is improved. We found that *Pipal* being very hardy tree is suitable for growing in dry areas. In this way *pipal* trees would supply green fodder and improve environment.

Shri Kulkarni told me about importance of *Umber* or wild fig tree. This tree is also called *Audumber*. The dried leaves and fruits of this tree are very useful as fodder. He suggested *Shivan (Gmelina arborea)* as another tree for fodder. The leaves and fruits of this plant are also liked by cattle. Besides, the wood of this tree is useful for preparing furniture. Shri Kulkarni told us that the leaves and fruits of banayan tree could also be used as fodder. Banyan tree has a long life and its plantation would also improve environment. There are large number of *Mahuva* trees all over the country. The flowers of this tree are very sweet and liked by the animals. The seeds of *Mahuva* are used for non-edible oils. Prof. Kulkarni told us that *Gulwel* is used in Ayurveda'. Cuttings of *Gulwel* vine mixed with little rice bran and salt could be used as a feed of cattle. A vine of wild bean locally called as *Abhai* is very useful as fodder. This plant is also drought tolerant and its leaves are available during dry months. The horses like the leaves of this plant. The wild tamarind tree has thorns on its branches. The pods of this tree could be used as fodder. There is one shrub which is locally known as *Kolista*. The seeds of this plant are called *Talimkhana*. These seeds could be used as feed to cattle.

Shri T.A. Kulkarni suggested that all the cultivable lands in our country would be necessary for food production while the waste lands could be utilised for growing fodder trees. He also suggested that the fodder trees could be planted along the border of fields so that food and fodder could be produced simultaneously.

Shri Kulkarni told me that he learned many practical things from shepherds. When he was staying at Nasik he reared milch buffaloes. However, the rate of mortality of calves was very high. After the death of calf the buffaloes reduced their milk yields. He tried various treatment on calves but could not succeed. There was a pasture adjacent to his house where shepherds were bringing their buffaloes along with their calves for grazing. He asked them why their calves are healthy and strong while his calves die few days after birth. However, the shepherds did not disclose their secret. Shri Kulkarni told me that he did not leave his attempt and continued his friendship with the shepherds. After some days the shepherds told him the technique of rearing buffalo calves. They told that the calf of the cow has the natural instinct of drinking required quantity of mother's milk. Besides, it starts eating grass from next day. The grass contains fibre and minerals which helps digestion. Contrary to this, the buffalo calf has no such instinct. It drinks excessive quantity of milk from its mother. Besides it does not eat grass. This results into indigestion.

Many a times it is infected with worms. These are the main reasons for mortality of buffalo calves. The shepherds told him that they put tender leaves of bitter gourd in the jaw of the calf. The calf starts chewing the leaves and develops habit of eating grass. This helps in digestion and also prevent infection of worms. Shri Kulkarni followed this method and found that the mortality of the buffalo calves was greatly reduced.

Shri Kulkarni has studied cultivation of fibre crops. He used to say that the wild variety of *Bhendi* (Okra) which grows on wastelands contains good quality of fibre. He grew jute crop and prepared cloth by mixing its fibre with cotton. He used to show the sample of this cloth to the visitors. The Indian Council of Agricultural Research and Central Jute Development Council had appreciated the research work of Shri Kulkarni. We prepared an experimental plot of jute crop under the guidance of Shri Kulkarni. The farmers who saw this crop were greatly impressed to see that jute crop could be grown under agro-climatic conditions of Konkan. Later we took up seed production of jute.

Shri Kulkarni gave us information about cultivation of *Agave* which is also an important fibre plant. During Pre Independence days a scientist from Germany made experimental plantation on Powai Hills near Bombay. Number of articles could be prepared from fibre of this plant. Shri Kulkarni was of the opinion that *Agave* plantations should be undertaken on wastelands in Konkan region. This would generate gainful employment in fibre industry. Even though Shri Kulkarni was educationist he had great interest in agriculture. I was greatly benefited by his rich experiences. It gave me new direction to conduct research in areas of fodder and fibre crops.

Shrimati Tarabai Modak

I received guidance from Shrimati Tarabai Modak for over sixteen years. She was the pioneer of child education in India. Tarabai earned her B.A. degree from University of Bombay in the early part of the century when very few women went for higher education. She was appointed Principal of Government Training College for Women at Rajkot in Gujarat. The Government gave her all facilities, including good salary, bunglow and a car. However, she left these facilities and decided to devote herself to the cause of child education at Bhavnagar. She worked for over twelve years in an Institution known as 'Dakshinamurti' and experimented in child education with her colleague Shri Gijubhai Badheka. She started 'Nutan Bal Shikshan Sangh' (New Child Education Organization) for promotion of child education. A journal called *Shikshan Patrika* (Educational journal) was published under the auspices of this organization. This journal spread the message of child education in the country. After working for several years in Gujarat she returned to Bombay and started an Institution called 'Sishu-Vihar' (Kinder Garten). She also started a training centre for training of teachers in child education. This was the first institution of its kind in Bombay State.

One day a thought came to Tarabai's mind that she had so far concentrated on child education work in cities and towns. However, India is a country of villages. Unless the message of child education is taken to villages there is no hope for development of children from villages. She therefore felt a need for starting child education work in villages.

Once she went to Panchagani to meet Mahatma Gandhi. She discussed her experience about child education with Gandhiji, who appreciated her efforts. However, he desired that her work should spread to the villages. He also desired that the educational tools used must be cheap and within the reach of poor children. He suggested that these tools should be prepared by rural artisans. He told her that India would soon attain Independence and would need a system of child education. He told her to accept this as a challenge and gave his blessings for her mission.

After meeting Mahatma Gandhi she returned to Bombay with a firm conviction that she would work in future in villages only. Accordingly, she selected Bordi (a coastal village in Thane district) and started an institution called 'Gram Bal Shiksha Kendra' (Rural Child Education Centre). The Centre was inaugurated on December 24, 1945 by Shri Balasaheb Kher, Chief Minister of Bombay State. Shri Kher in his address said "Tarabai came to this village with the blessings of Mahatma Gandhi. By starting this centre she is sowing the seed of child education. This seed will grow into a big tree and spread all over India".

Bordi was the *Karmabhoomi* (a place of worship through dedicated service and selfless action) of Acharya Bhise and the educational atmosphere which existed here was congenial, for the new institution of Tarabai Modak. The local citizens also extended all the help and cooperation in her work. They included Shri Madhavrao Raut and his brother Shri Madanrao, Shri Atmarampant Sawe, Shri Mukundrao Sawe, Dr. Mohanlal Pathak, Shri Shamrao Patil and Shri Thakorbhai Shah.

Smt. Tarabai Modak was very fortunate to have Anutai Wagh as her colleague who also devoted her life for child education. Both of them did experimented with child education. The children's school was quite new to Bordi village. Smt. Anutai Wagh used to take children in small processions. The children were singing songs and ringing bells. A bullock cart followed the procession to carry children who were unable to walk long distances. Children of all religions and communities participated in the procession. The parents also developed interest in child education. This was the first experiment of organising child education in rural areas and Tarabai Modak made great efforts to make it a success.

Once Shri Jugatram Dave a great Gandhian worker from Gujarat visited the Gram Bal Shiksha Kendra at Bordi. He told Tarabai "there are orchards of coconuts, mangoes, chikooes and pomeloes in this area. They are locally called *Wadi*. The hamlets in the village are named as Chafawadi, Amritwadi etc. Why don't you also name your children's school as *Balwadi* (*Bal*=child). Smt. Tarabai Modak liked this word and decided to name her children's school as *Balwadi*.

There is a tribal population in surrounding hamlets of Bordi. Smt. Tarabai Modak started a school for tribal children in a hamlet called Darjiwadi. The school was held in the front courtyard of a house, which is locally called *Angan*. Tarabai named her children school *Anganwadi*. Thus the names *Balwadi* and *Anganwadi* originated from institutions of Tarabai Modak at Bordi. Both these names have become popular throughout India.

Shrimati Tarabai Modak carried out experiments for twelve years in her Gram Bal Shiksha Kendra at Bordi. She made necessary modifications in the Montessory system of child education suitable to the local environment. While doing this she took care that the original principles of Montessory system were not disturbed. She developed educational tools which were cheap and could be prepared by teachers and village artisans. She prepared them from local material like newspapers, bamboo baskets, empty tins and bottles. She placed emphasis on recycling of wastes for preparing educational tools.

Shrimati Tarabai Modak started a teachers training centre to impart training in child education. Teachers from all over India were attending courses at this Centre. Four languages viz. Marathi, Gujarati, Hindi and English had to be used. It was a difficult task but Shrimati Tarabai Modak took the responsibility as she wanted to spread the message of child education throughout the country. The teachers who were trained from this centre started *Balwadis* in rural areas of India. In this way the Gram Bal Shiksha Kendra at Bordi played an important role in spreading child education in India.

The surrounding areas of Bordi were inhabited by tribals. Shrimati Tarabai Modak found that the problems of educating tribal children were quite different and could not be solved simply by starting *Balwadis*. An integrated approach would be necessary. She therefore decided to work in tribal areas, and selected Kosbad Hill a small vilage which was inhabited by tribals. She started a project called 'Vikas wadi' (development garden) in 1957. The main objective of the project was to evolve an integrated approach for child education among the weaker sections like tribals. It consisted of a creche centre, *balwadi* and primary school at one place. This was the first experiment of its kind in India.

Taking into consideration the importance of this experiment Shrimati Tarabai Modak organized an All India Child Education Conference at Kosbad Hill in 1957 to know the views of educational experts and social scientists from India. The conference was presided by well known educationist and Gandhian thinker Shri Kakasaheb Kalelkar. All the participants appreciated the integrated approach of child education developed by Shrimati Tarabai Modak. The conference also passed a resolution requesting the Government to start pre-school child education in the country.

Smt. Tarabai started a centre for training of teachers for *Balwadi* and primary school. The project of Vikas Wadi (viz. integrating creche, balwadi, primary school and training centre) received the support of Ministry of Education, Government of India. However, donations were required for purchase of land and construction of building. Shrimati Tarabai Modak used to go to Bombay and Ahmedabad for collection of funds. It took about one year to collect the funds required for the project. The foundation stone of the building of Vikaswadi project was laid by Shrimati Shantadevi Gaikwad of Baroda.

A land measuring 15 acres was purchased for setting up a training and demonstration farm. Shrimati Tarabai gave me the responsibility of developing this farm. A well was constructed and a pumping set was installed. Various fruit trees like mangoes, chikooes, coconuts, guavas, papayas and bananas were planted. Similarly vegetables were grown as intercrops. Fields were laid for growing rice. All the produce from the land was utilised for feeding the children from creche, *balwadi* and primary school.

The construction of buildings took about two years. Till then Shrimati Tarabai Modak and her colleagues were staying in the building of the Kosbad Institute. Since I was their neighbour I got the opportunity to study the integrated approach of child education. I found that this is a right approach for promotion of education among the weaker sections.

The agricultural farm attached to Vikaswadi was utilized for imparting training to the teachers of *Balwadi* and primary school. Shrimati Tarabai used to tell me that if the teachers learn the skills of growing vegetables and fruit trees they could motivate the parents of the children to grow them. This would help to improve the nutrition of children.

When Smt. Tarabai Modak started a primary school, all the children from the village were not attending. She used to go from house to house to know the reason for their absence. She found that the children had to go to meadows for grazing their animals. Tarabai thought that instead of calling the children to the school why couldn't teachers go to the meadows to conduct classes. Accordingly, she started a 'meadows school'. It was a pathbreaking experiment of open education for primary school children. Ministry of Education, Government of India published a book viz., 'Meadows school' based on this experiment.

The tribal women usually worked at nearby orchards or at collection of fuelwood from the forest. They had to leave their infants at home. Naturally, the grown up children had to take care of infants and were, therefore, unable to attend the *Balwadi*. Smt. Tarabai attached a creche centre to the *Balwadi*. The children were, then, able to keep the infants in the creche and attend the *Balwadi*.

My wife, Meena, was greatly inspired by the work of Smt. Tarabai Modak. She completed a one-year course on pre-school education under the guidance of Smt. Modak. After completion of this course she started two *balwadis* on the farms of the Kosbad Institute and Ashram School at Waki. Creche centres were attached to these *balwadis*. These freed the working women to work on the farm, while their children were able to attend the *Balwadi*. We started two more creche centres and *balwadis* at Chimbave and Manpada villages. We found that creche centres and *balwadis* were a boon to both children and working women.

Smt. Tarabai Modak told me that the children from the weaker sections were unable to attend primary school due to economic compulsions. Many children had to go for work, since parents perceived them as income generating assets. Smt. Tarabai used to suggest that we find part time employment for such children. Accordingly we were giving some light work like collection of eggs, feeding chicken and harvesting vegetables for about two hours in the morning. The children used to attend the school in the afternoon. Due to such part time work the attendance in the school improved considerably.

Smt. Tarabai started cottage industries in her Vikaswadi which enabled the children to earn some money. The educational tools required in *balwadis* were prepared by these children in carpentry workshop. A printing press was also started to train these children in printing.

The Government of India recognised the services of Smt. Tarabai Modak and awarded her "Padma Bhushan". Smt. Indira Gandhi, Prime Minister of India visited Vikaswadi and was greatly impressed to see her work. Shrimati Tarabai Modak considered child education as her mission of life. She was therefore known as 'Montessary of India'.

Shri Shamrao Patil

My father Shri Shamrao Patil was associated with the freedom struggle started by Mahatma Gandhi since 1920. Swami Anand, a close associate of Mahatma Gandhi while travelling from Bombay to Surat saw the Arabian Sea near Gholvad Railway station. He got down at the station and came to Bordi. My father was the first person to meet him. Swamiji was dressed in Khadi, and so my father thought him to be Gandhiji's man. Swamiji told him that Gandhiji's son Ramdas was ill and had been advised by the doctors to stay near the sea. He therefore, liked the atmosphere of Bordi and had decided to bring Ramdas here for convalascence. My father came in close contact with Swami Anand and became his disciple. Rashtramata Kasturba Gandhi also came to stay at Bordi alongwith her son Ramdas. Swami Anand brought Kasturba to our house. Many Gandhians like Narhari Parikh, Jugat Ram Dave also often visited our house. My father used to tell me about their contribution in India's freedom struggle. Kaka Saheb Kalelkar another colleague of Mahatma Gandhi also came to Bordi in 1929. My father used to meet him regularly as a volunteer. Kakasaheb was very happy with the pleasant nature of my father and used to call him 'Priya Darshan Shamrao.' My father requested Kakasaheb to give him an opportunity to participate actively in the freedom struggle.

Kakasaheb told him he would definitely consider my father's request provided he was prepared to become a full time worker in the freedom movement. He also asked my father that since he was an only son he should seek consent from his mother as well as his wife. My father was less than two years old when his father expired and so my grand mother was dependent on her only son. However, when my father told his mother that Shri Kakasaheb Kalelkar is considering to give him an opportunity to participate in the freedom struggle led by Mahatma Gandhi, my grand mother gave her consent gladly. At that time, I was about two and half years old while my younger brother Sharad was only six months old.

Shri Kakasaheb Kalelkar sent a telegram to my father stating that he should immediately come to Ahmedabad to attend a camp at Gujarat Vidyapeeth. This camp was organized in preparation for Gandhiji's Dandi March to initiate the Salt Satyagrah (civil disobedience). My father attended this camp where Mahatma Gandhi addressed the participants every day. The speeches of Mahatma Gandhi considerably inspired my father and he took a pledge to commit himself to the freedom movement. My father participated in the Salt Satyagrah offered by Shri Mahadeo Desai at Ahmedabad. Mahadevbahai was secretary to Gandhiji. Subsequently, my father went to Dharasana alongwith a team of doctors to treat the satyagrahis (volunteers) brutally beaten by the police. He then organized a Salt Satyagrah all over Thane District. In our village Bordi he led a procession to the Arabian Sea for the preparation of salt.

Gandhiji's colleague Swami Anand established Gandhi Ashram at Bordi in 1931. The President of Karachi Congress, Sardar Vallabh Bhai Patel inaugurated this Ashram. Shri Rajendra Prasad presided over the occasion. My father joined as an inmate in this Ashram. It was Swami Anand who inspired my father to take up political work for achieving freedom of the country. When my father was in the Gandhi Ashram, I was only four years old. The inmates of the Ashram used to organize morning and evening processions for awakening people to the freedom struggle. They sang national songs all of which I learnt. My father recited to me stories of martyrs who had laid down their lives for India's freedom struggle.

During the Non-Cooperation Movement the Gandhi Ashram at Bordi was confiscated. My father was arrested and was imprisoned for sixteen months in the Visapur Prison in Ahmed Nagar District. During this period he suffered from typhoid for forty two days. However, due to God's grace he recovered from this illness. During this time Shri Ramlabhai Machhi from Dehri village, a co-prisoner of my father's rendered valuable services to him. Similarly, Dr. Dinanath Churi, a colleague of my father from Bordi used to visit Visapur prison and offered medical help to him. My father also participated in the Quit India Movement of 1942 and was imprisoned for about twelve months in the Thana Central Jail.

My father believed that untouchability was a great blot on our country. He, therefore, took up a programme for its removal which he initiated in his home itself. He invited dalits and took meals with them. His mother imparted considerable support to him in this mission. She treated the dalits as her sons. One dalit woman Smt. Dhanabhai Surti even helped my grand mother and mother in cooking food. As a result of this the orthodox Hindus boycotted my father and other members of our family. However my father was an ardent follower of Gandhiji's path of self-restraint. For about two years my father and all our family members voluntarily chose to restrict themselves from attending any religious or social functions. This had the implied effect on the orthodox locals who realised their mistake. From the very childhood I have been inspired by my father's concern for the plight of dalits and hence a spirit of service for the dalits has inculcated in me.

My father also used to wrote among the surrounding tribals. The landlords in our area were forcibly encroaching upon the lands of the tribal tenants. My father helped the tribals in this regard. Similarly, during the spread of epidemics like small pox, cholera etc. he used to render medical help as he himself was a registered medical practitioner. My father often told me about the plight of the tribals and the efforts he was making to render some service to them. These attributes of my father's have gone a long way to awaken social consciousness in me.

My father became a full time worker in the freedom movement due to the strong support offered by my grandmother and mother. We had a small piece of land in which rice was grown. My grandmother and mother did all the manual work like sowing, transplanting, harvesting etc. We were self-sufficient in foodgrains. However, we did not have any cash income. So my grandmother and mother used to bring coconut husk and extracted coir out of it. It was through the sale of this coir that they earned some cash income. However, both of them gladly accepted hardships and encouraged my father to participate in India's freedom movement.

In 1936, my father constructed a small well on our land. The neighbouring farmers gave him some coconut and chikoo saplings. He raised a small orchard on this land. We also raised vegetables in our orchard. I joined my grandmother and mother in operations like manuring, watering, harvesting etc. in the orchard. This is how I acquired practical knowledge in agriculture. This orchard helped us get some cash income through the sale of fruits and vegetables. However, we maintained a simple way of living due to limited income, another attribute I inculcated from early childhood.

My father was the President of the Thane District Congress Committee for several years. At that time there was no office for the Congress Committee. Our house was used as the office and my father was used to writing letters in his own handwriting. He used to give me the envelopes for writing the addresses of workers in the district, which helped me to know all the social workers in Thane district. My father always attended the sessions of the All India Congress Committee. On his return he used to tell me about the great leaders like Mahatma Gandhi, Jawahar Lal Nehru, Sardar Patel and Dr. Rajendra Prasad. This has greatly enhanced my knowledge of the history of the freedom movement. My father was appointed the General Secretary of Maharashtra Pradesh Congress Committee. He had to stay in the Congress House at Pune. At that time no honorarium was paid to the office bearers of the Congress

Committee and my father thus had to lead a simple life. I was studying in the Agricultural College in Pune and many a times my father used to come to take his meals in the canteen of our hostel.

The leaders of the Congress Party requested my father to seek election for the Bombay Legislative Assembly in 1952. At that time there was no practice of asking for the Congress ticket. The sacrifice of the candidate in the freedom movement as well as his deep commitment in the party organization was taken into consideration. My father was elected with a good majority in that election. At that time legislation like the Tenancy Act was passed by the Bombay Legislative Assembly for initiating land reforms. My father devoted himself full time to see that land reforms were implemented effectively. My father was again given a Congress ticket in 1957. At that time political movement for Sanyukta Maharashtra (United Maharashtra) was in full swing. However, due to my father's commitment to selfless service, he was the only congress candidate to be elected from three coastal districts. Shri Yashwant Rao Chavan who was the Chief Minister of the bigger bilingual Bombay state included my father in his Cabinet. He was given the portfolios of Panchayats, Sarvodaya, Forest, Labour, Cooperative Societies and Salt Lands Development Boards. During his tenure as Minister he continued to lead a simple life. He travelled in third class even though he was eligible for the higher class. Whenever he visited our village at Bordi he travelled in a horse tonga from Gholvad railway station to Bordi. He never availed for a Government car to come from Thane to Bordi. My father was elected in 1962 to Bombay Legislative Assembly for the third time. In this way he was Member of the Assembly for fifteen continuous years.

My father worked from 1920 to 1947 (27 years), for India's freedom movement during which time he spent 29 months in jail. He underwent great hardship during this period. However, he did not accept pension or any benefits eligible for the freedom fighters. He considered it his sacred duty to free Mother India from the clutches of the foreign rule and devoted himself to the cause of tribal upliftment. The landlords in our area were evicting tribal tenants due to their ignorance. On such occasions my father helped the tribals fight for restoration of their land. He often took such cases to the Chief Minister also. He even donated about two and a half acres of his land as *Bhoodan* (land grant) to a tribal family.

My father told me that the All India Congress Committee had appointed a National Planning Committee in 1938 under the Chairmanship of Pandit Jawahar Lal Nehru. This Committee in its report laid great emphasis on the development of agriculture in India. When I passed my matriculation my father suggested me to take higher education in agriculture. Accordingly, I entered Agricultural College at Pune and earned my Bachelors degree from University of Bombay in 1949. It was because of his vision that I got an opportunity to work in the field of agriculture and rural development.

When I was working at the Kosbad Institute he told me to commit myself to the service of the poor tribals. Even during his old age he kept me free from home responsibilities which helped me to devote my entire time for the service of the tribals.

On 10th June 1975, my father had a severe heart attack . The doctors tried their best, but they did not succeed. About 5 O'clock in the morning his body became very cold and one of our relatives suggested that we should apply brandy to warm the body. Even though he was in a very critical condition he heard this and told us not to touch brandy to his body. After few minutes he breathed his last, but did not allow liquor to be touched to his body. My father tried to practice the teachings of Mahatma Gandhi and he was of the opinion that Gandhiji's teachings would be relevant foi many years to come.

Shri Khanderao Save

My father-in-law Shri Khanderao Save guided me in my work for tribal upliftment. Shri Save received his education under Acharya Bhise and thus led a principled life. During the British rule Shri Save was the Backward Class Officer of Thane and Colaba Districts. At that time the Social Welfare Department was called Backward Class Department. The forest contractors in Thane and Colaba districts did not pay proper wages to the tribals. They used to exploit the tribals and those who resisted were greatly harassed. The Government of Bombay passed an Act to pay proper wages to the tribals working in the forest areas. Shri Save organized the tribals and explained to them about the provisions of the Act. He prosecuted many forest contractors who were not paying proper wages and exploiting the tribals. The forest contractors organized themselves and made representation for transfer of Shri Save. However, their request was not paid any heed and Mr. Save remained in the same place. The Government was aware that Shri Save was the most competent officer and person of great integrity.

While working among the tribals Shri Save studied socio-economic conditions of Warli tribe and wrote a thesis on it. He received the guidance of Dr. G.S. Ghuraye, the well known Sociologist in this study. He submitted his thesis to the University of Bombay for which he was awarded the M.A. Degree. This thesis was published in a book called

Warlis in 1945. Shri Balasaheb Kher, Chief Minister of Bombay State had written a foreword for this book. This book is used as a reference manual in tribal literature.

After attainment of independence in 1947 the Chief Minister of Bombay State, Shri Balasaheb Kher appointed Shri Save as the Deputy Collector and Special Tribal Protection Officer at Dahanu. He was also promoted in the cadre of the Indian Administrative Service. The Tenancy Act was passed around that time. Shri Save organized the tribals for the implementation of the Tenancy Act. He spoke the tribal dialect and so the tribals had great attachment for him. It was because of the sincere efforts of Shri Save that the tribals became owners of their land. The contribution made by Shri Save was appreciated by great persons like Shri Thakkar Bappa who was President of Bhartiya Adimjati Sevak Sangh (All India Tribal Service Organization). He himself came to the house of Shri Save at Bordi to give his compliments. The Chief Minister of Bombay Shri Balasaheb Kher and Home Minister Shri Morarji Desai also appreciated Shri Save's efforts in curbing exploitation of tribals.

Shri Save had worked for over twenty five years with the tribals and had studied their social and economic conditions in great depth. He imparted to me considerable guidance in my tribal development work. He explained that due to continued exploitation, the tribals had lost their self-esteem and confidence amongst themselves. This could be created only by education and through spreading literacy. However, it would take two or three generations of strong external initiative to achieve this goal. Thereafter the tribals would be able to stand independently.

He told me that the responsibility of tribal development work should not be left only to the tribals. Non-tribals should also devote themselves for tribal upliftment work. He quoted examples of Shri Thakkar Bappa, Acharya Bhise, Shri Jugatram Dave, Shri Lakshmikant Shrikant who devoted their lives for the upliftment of the tribals. Similarly, Dr. Albert Switzer and Mr. David Livingstone who did life long service to the tribals in Africa were non-tribals. He felt that young people in our country should consider tribal upliftment as a humanitarian work and devote themselves for this noble cause.

When I asked about his advice about the working of the non-government welfare organizations he quoted the example of Shri Thakkar Bappa. Thakkar Bappa used to say that welfare organizations should always be in debt and the workers of such organizations should take a begging bowl in their hand and move among the people for collecting funds. This alone would impart accountability to people. Shri Save told me another example about Shri Anna Saheb Karve who was a great reformer for the upliftment of women. He was running one educational institution for women at Hingane near Pune. He used to move from house to house for collection of funds in the city of Pune. When Shri Save was a Backward Class Officer in Pune, Anna Saheb Karve came to his house. Shri Save told him that he was a very small officer and would not be able to pay a large amount. Shri Karve immediately replied that he accepted even half a rupee. He further told him he has not come simply for funds but also to create friendship and goodwill among the people. Shri Save gave him a rupee but thereafter Shri Save went to Hingane regularly and donated to the institute of Shri Anna Saheb Karve. This shows that if the workers of the institutions go to the people regularly people would definitely respond. Shri Save told me that the real spirit of service in institutions would be created if they are 'people oriented'.

Shri Save also informed me about the memoirs of Dr. Veriar Elvin. He was an Englishman and worked life-long among the tribals of Assam North-East and Madhya Pradesh. He had written several books on the different tribes of India. Shri Save told me that Mr. Elvin came to Mokhada block which was one of the first tribal development blocks in the state of Maharashtra chosen by the Government of India. Mr. Elvin told Mr. Save that together with their economic development the tribal culture should also be maintained. Shri Save advised me that in our Ashram School and tribal hostels in Kosbad we should encourage boys and girls to preserve their folk songs, folk dances and tribal paintings. We listened to the advice of Shri Save and encouraged our students accordingly.

16

RECOGNITION

Recognition and Opportunities to Work in Wider Fields

In 1978 the Jamnalal Bajaj Foundation instituted two awards of Rs. 1 lakh each in memory of Late Shri Jamnalal Bajaj who was a close associate of Mahatma Gandhi. One Award was meant for the outstanding contribution done in the field of Social work and the other was for application of Science and Technology for rural development. Justice Hidayatullah, Vice President of India was the Chairman of the Council of Advisers of the Foundation.

He announced in a Press Conference at Bombay in August 1979 that Dr. Jayant Patil had been selected for the Jamnalal Bajaj Award for the Application of Science and Technology for Rural Development. Dr. Homi Sethna, Chairman of the Atomic Energy Commission of India was also the Chairman of the Selection Committee for this Award. Dr. M.S. Swaminathan, the then Director General of Indian Council of Agricultural Research had recommended my name. I received this Award on 4th November, 1979 at New Delhi at the hands of Shri Neelam Sanjiva Reddy, President of India.

Shri Vijay Tendulkar, a well known Dramatist interviewed me for one hour in the programme entitled *Pratibha and Pratima* on Bombay Doordarshan. This Award gave my work in rural reconstruction recognition at the National level.

The Konkan Agricultural University at Dapoli conferred on me an honorary degree of Doctor of Science in 1981. Dr. O.P. Gautam, Director General, Indian Council of Agricultural Research was the Chief Guest at the Convocation. Dr. Gautam gave compliments for my contribution in the field of agriculture and rural reconstruction.

In 1981 Shri A.R. Antulay, Chief Minister of Maharashtra appointed me as Honorary Adviser to the Government of Maharashtra in Horticulture. I worked in this post for about ten years. While working in this post my experiences at Kosbad Institute were very helpful in developing the newly created department. I took up the programme of establishment of horticultural nurseries at district level for supply of quality planting material to the farmers, utilizing wastelands for horticultural development, dryland horticulture for farmers whose lands were dependent on rains, plantation of mangoes and cashews for the farmers of hilly areas. I evolved a pattern of 'companion cropping' which consisted of growing fruit trees alongwith food crops. This pattern would not affect the production of food crops. I found that horticulture could not only provide cash income to the small farmers and tribals, but also help to enrich the diet of family through consumption of fruits.

Shri Rajiv Gandhi, Prime Minister of India established a National Wasteland Development Board in 1985. He gave emphasis on growing fuelwood and fodders in these wastelands. Two committees were constituted by the Board for that purpose. I was invited to work on the Committee of Grasses and Fodders. While working on this Committee my experiences in fodder development at the Kosbad Institute were found to be very useful. The Committee recommended the introduction of nutritious fodders like *Stylosanthes hamata, Stylosanthes scabra* and *Siratro (Macroptelium atroperpurium)* in the wastelands of India. The Committee also recommended setting up of seed farms to supply seeds of these new fodders. I visited wasteland development programmes in the country and found that both Government and non-Government Organizations had introduced these fodders.

In 1987 Shri Shankarrao Chavan, Chief Minister of Maharashtra appointed me Chairman of the Maharashtra State Khadi and Village Industries Board. I worked in this post for a period of four and a half years. While working in this post, I tried to introduce sericulture in Maharashtra. Maharashtra was non-traditional State and hence we gave emphasis on creating infrastructure like training of farmers, establishment of grainages for supply of disease free layings (DFL), procurement of cocoons, reeling and weaving centres. We also took up the cluster approach which facilitated implementation of the programme effectively. Due to this approach Maharashtra has now emerged as the leading State in sericulture among the non-traditional States.

I strengthened the programme of bee-keeping. The Western Ghats as well as the Satpura Hills due to their forests, were rich in flora. We have promoted Bee-Keepers' Cooperatives in these areas. We found that the tribals and landless people from hills get good income through bee-keeping. I took up a programme of fruit processing. The Ratnagiri and

Sindugadh districts of Konkan area were famous for alphanso mangoes. Educated youth from these areas were encouraged to set up mango processing units. The processed mangoes were in great demand both in domestic and international markets. There were large numbers of Mahila Mandals (Women's Organizations) in the State of Maharashtra. We gave them support in processing of cereals and pulses and also in the preparation of pickles and papad making. We also helped them in marketing of their produce. Large numbers of educated youths in the rural areas were unemployed. We arranged for their technical training and gave support to set up small workshops in carpentry and smithy. These workshops catered to the needs of villages in preparation of housing material, repair of agricultural machinery and light vehicles. These workshops were helpful in generation of gainful employment for the rural youths.

We undertook a programme of *Amber Charkhas* in Wardha district and found that this programme gave employment to large number of rural women. While working in the Maharashtra State Khadi and Village Board I tried to strengthen the 'Artisan's Cooperatives' in the State. These Cooperatives helped in organising credit to the artisans and also in marketing of their produce. I found that setting up Artisans Cooperatives would strengthen the village industries in India.

Shri Sudhakarrao Naik, Chief Minister of Maharashtra continued my appointment as Chairman, Khadi and Village Industries Board and also Honorary Adviser in Horticulture. Shri Naik created an independent Water Conservation Department in Maharashtra Government which helped to strengthen the programme of horticulture in rainfed and dry areas. Shri Naik entrusted me with the responsibility of Secretary of Vasantrao Naik Agricultural Research and Rural Development Foundation. While working in this post, I received valuable experience of agricultural research and rural development activities in the State.

The Prime Minister of India Shri P.V. Narasimha Rao appointed me as Member of the Planning Commission in August 1991. I was given the portfolios of Agriculture, Rural Development, Irrigation, Cooperation and Panchayati Raj. While working in the Planning Commission I got an opportunity to prepare the Eighth Five Year Plan. While preparing this Plan, the experiences which I gained during the last fourty years in agriculture and rural reconstruction were very useful. During the Eighth Five Year Plan thrust was given on diversification of agriculture through horticulture and sericulture. Since I had worked in both these areas I was able to make significant contribution in preparing various programmes. The Eighth Plan also gave emphasis on strengthening animal husbandry programmes. Fodder being one of the important components for improving the animal productivity, I was able to contribute in that area too. Since I had worked for the development of tribals I was able to make contribution in preparing Plan for upliftment of weaker sections and strengthening the social sector. While working in the Planning Commission I realised that field experiences were essential and of considerable help in knowing the problems at grassroot levels.

In my rural reconstruction work I received tremendous cooperation and support both from Government and non-government organizations. I am very thankful to the organizations who recognized my work and bestowed honours on me. I express my deep sense of gratitude to the Prime Minister of India, Shri P.V. Narasimha Rao for giving me an opportunity to work as Member, Planning Commission which enabled me to utilize my grassroot experiences at national level.

INDEX